J. Paulo Davim
Editor

Tribology of Nanocomposites

Springer

Editor
J. Paulo Davim
Department of Mechanical Engineering
University of Aveiro
Campus Santiago
Aveiro
Portugal

ISSN 2195-0911 ISSN 2195-092X (electronic)
ISBN 978-3-642-33881-6 ISBN 978-3-642-33882-3 (eBook)
DOI 10.1007/978-3-642-33882-3
Springer Heidelberg New York Dordrecht London

Library of Congress Control Number: 2012952012

Printed on acid-free paper

Springer is part of Springer Science+Business Media (www.springer.com)

Preface

Today, it is usual define nanocomposite "*as a multiphase solid material where one of the phases has one, two or three dimensions of less than 100 nm or structures having nano-scale repeat distances between the different phases that make up the material*". Recently, the use of nanocomposites with polymer, metal or ceramic matrices has increased in various areas of engineering and technology due to their special properties, with applications in bioengineering, battery cathodes, automotives, sensors and computers, as well as other advanced industries.

Tribology is defined as "*the science and technology of interacting surfaces in relative motion*" and embraces the study of friction, wear and lubrication. Within the tribology literature of nanocomposites wear rate and friction coefficient are generally reduced with the addition of nanoscopic filler particles for all matrices with special emphasis in polymer matrices. Friction coefficients and wear rates are often discussed in nanocomposites tribology.

The Chap. 1 of this book provides tribology of bulk polymer nanocomposites and nanocomposite coatings. Chapter 2 is dedicated to nano- and micro-PTFE for surface lubrication of carbon fabric reinforced polyethersulphone composites. Chapter 3 described Tribology of MoS_2-based nanocomposites. Chapter 4 contains information on friction and wear of Al_2O_2-based composites with dispersed and agglomerated nanoparticles. Finally, Chap. 5 is dedicated to wear of multi-scale phase reinforced composites.

This present book can be used as a research book for final undergraduate engineering course or as a topic on materials at the postgraduate level. Also, this book can serve as a useful reference for academics, tribology and materials researchers, materials, physics and mechanical engineers, professional in nanocomposites and related industries. The interest of scientific in this book is evident for many important centres of the research, laboratories and universities as well as industry. Therefore, it is hoped this book will inspire and enthuse others to undertake research in this field of tribology of nanocomposites.

The Editor acknowledges Springer for this opportunity and for their enthusiastic and professional support. Finally, I would like to thank all the chapter authors for their availability for this work.

Aveiro, Portugal, August 2012 J. Paulo Davim

Preface

Contents

Contributors

M. D. Bermúdez Grupo de Ciencia de Materiales e Ingeniería Metalúrgica, Departamento de Ingeniería de Materiales y Fabricación, Universidad Politécnica de Cartagena, Campus de la Muralla del Mar, 30202 Cartagena, Spain, e-mail: mdolores.bermudez@upct.es

Jayashree Bijwe Industrial Tribology, Machine Dynamics and Maintenance. Engineering Centre (ITMMEC), Delhi, Indian Institute of Technology, Hauz Khas, New Delhi 110016, India, e-mail: jbijwe@gmail.com

F. J. Carrión Grupo de Ciencia de Materiales e Ingeniería Metalúrgica, Departamento de Ingeniería de Materiales y Fabricación, Universidad Politécnica de Cartagena, Campus de la Muralla del Mar, 30202 Cartagena, Spain

C. Espejo Grupo de Ciencia de Materiales e Ingeniería Metalúrgica, Departamento de Ingeniería de Materiales y Fabricación, Universidad Politécnica de Cartagena, Campus de la Muralla del Mar, 30202 Cartagena, Spain

Kunhong Hu Department of Chemistry and Materials Engineering, Hefei University, Hefei 230022, People's Republic of China, e-mail: hukunhong@163.com

Xianguo Hu Institute of Tribology, Hefei University of Technology, Hefei 230009, People's Republic of China, e-mail: xghu@hfut.edu.cn

Yang Jiang School of Materials Science and Engineering, Hefei University of Technology, Hefei 230009, People's Republic of China

Zhenyu Jiang Department of Engineering Mechanics, South China University of Technology, 381 Wushan Road, Guangzhou 510641, People's Republic of China, e-mail: zhenyujiang@scut.edu.cn

Jinjun Lu State Key Laboratory of Solid Lubrication, Lanzhou Institute of Chemical Physics, Chinese Academy of Sciences, Lanzhou 730000, People's Republic of China, e-mail: jjlu@licp.cas.cn

Junhu Meng State Key Laboratory of Solid Lubrication, Lanzhou Institute of Chemical Physics, Chinese Academy of Sciences, Lanzhou 730000, People's Republic of China

G. Ojados Grupo de Ciencia de Materiales e Ingeniería Metalúrgica, Departamento de Ingeniería de Materiales y Fabricación, Universidad Politécnica de Cartagena, Campus de la Muralla del Mar, 30202 Cartagena, Spain

J. Sanes Grupo de Ciencia de Materiales e Ingeniería Metalúrgica, Departamento de Ingeniería de Materiales y Fabricación, Universidad Politécnica de Cartagena, Campus de la Muralla del Mar, 30202 Cartagena, Spain

Jian Shang State Key Laboratory of Solid Lubrication, Lanzhou Institute of Chemical Physics, Chinese Academy of Sciences, Lanzhou 730000, People's Republic of China

Mohit Sharma Leibniz-Institut für Polymerforschung Dresden e.V., Hohe Strasse 6, 01069 Dresden, Germany

Xiaojun Sun State Key Laboratory of Solid Lubrication, Lanzhou Institute of Chemical Physics, Chinese Academy of Sciences, Lanzhou 730000, People's Republic of China

Tao Wang State Key Laboratory of Solid Lubrication, Lanzhou Institute of Chemical Physics, Chinese Academy of Sciences, Lanzhou 730000, People's Republic of China

Yufu Xu Institute of Tribology, Hefei University of Technology, Hefei 230009, People's Republic of China

Zhong Zhang National Center for Nanoscience and Technology, No.11 Zhongguancun Beiyitiao, Beijing 100190, People's Republic of China

Tribology of Bulk Polymer Nanocomposites and Nanocomposite Coatings

M. D. Bermúdez, F. J. Carrión, C. Espejo, J. Sanes and G. Ojados

Abstract Polymer-matrix nanocomposites are increasingly been applied as structural materials in the aerospace, automotive and chemical industries, in a number of tribological applications including such components as bearings, gears, cams, seals, vacuum pumps and in biological or medical applications in prothesis and implants, particularly in areas where traditional fluid lubricants can not be used. The base polymers studied have now extended from the original high temperature polymers to a wide range of thermoplastics and thermosets. In the present review we will focus on recent developments on polymer nanocomposites tribology including nanoparticles, with an emphasis on polymer nanocomposites containing carbon nanotubes, which are experiencing a large expansion and, as the tribological performance is mainly a function of surface properties, we will consider not only bulk nanocomposites, but also the development of polymer matrix nanocomposite coatings, which present very promising applications in the reduction of friction and wear.

1 Introduction

1.1 Polymer Matrix Nanocomposites in Tribology

The 1995 review on the tribology of polymer composites by Friedrich et al. [1], contained 37 references, mainly on friction and wear of high temperature polymers such as polyetheretherketone (PEEK) and polytetrafluoroethylene (PTFE)

Funding agency: MINECO (Spain). Grant # MAT2011-23162

M. D. Bermúdez (✉) · F. J. Carrión · C. Espejo · J. Sanes · G. Ojados
Grupo de Ciencia de Materiales e Ingeniería Metalúrgica, Departamento de Ingeniería de Materiales y Fabricación, Universidad Politécnica de Cartagena,
Campus de la Muralla del Mar, 30202. Cartagena, Spain.
e-mail: mdolores.bermudez@upct.es

J. Paulo Davim (ed.), *Tribology of Nanocomposites*, Materials Forming, Machining and Tribology, DOI: 10.1007/978-3-642-33882-3_1,

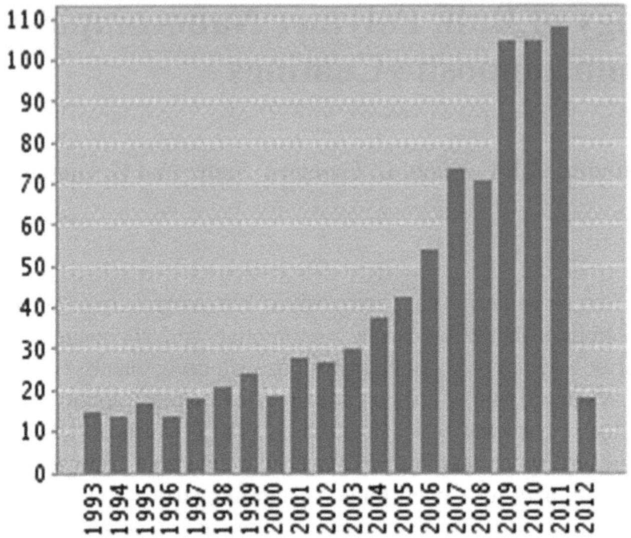

Fig. 1 Evolution of the number of research papers on tribology of nanocomposites (ISI Web of Knowledge © Thomson Reuters; February 2012)

modified by the addition of internal lubricants or fibre reinforcements. That is, no polymer nanocomposites were included. The field of nanocomposites tribology has grown and developed over the past two decades. The interest in the field, the number of research groups and the corresponding publications has been growing steadily, particularly since the year 2000.

The evolution of the number of scientific papers on friction, wear and lubrication of nanocomposites and their corresponding citations, according to the ISI Web of Knowledge database are represented in Figs. 1 and 2, respectively.

Polymer-matrix nanocomposites are increasingly been applied as structural materials [2] in the aerospace, automotive and chemical industries, in a number of tribological applications including such components as bearings, gears, cams, seals, vacuum pumps and in biological or medical applications in prosthesis and implants, particularly in areas where traditional fluid lubricants can not be used [3]. The base polymers studied have now extended from the original high temperature polymers to a wide range of thermoplastics and thermosets.

In a previous review [4], we have focused on the friction and wear reduction of polymers by the addition of nanoparticles, mainly metal oxides.

In the present review we will focus on recent developments on polymer nanocomposites tribology including nanoparticles, with an emphasis on polymer nanocomposites containing carbon nanotubes, which are experiencing a large expansion and, as the tribological performance is mainly a function of surface properties, we will consider not only bulk nanocomposites, but also the

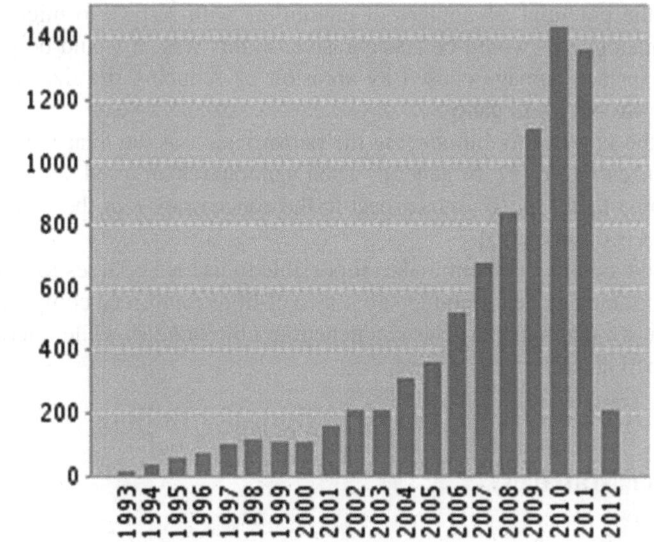

Fig. 2 Citations per year corresponding to the papers on tribology of nanocomposites represented in Fig. 1 (ISI Web of Knowledge © Thomson Reuters; February 2012)

development of polymer matrix nanocomposite coatings, which present very promising applications in the reduction of friction and wear.

1.2 Mechanisms for the Tribological Improvement of Polymers by Nanophases

The main principles which have been proposed to explain the tribological improvement of polymers by means of nanophases can be summarized in the following mechanisms:

1. Increase of the mechanical resistance and thermal stability of the polymer-matrix nanocomposite with respect to the base polymer.
2. Formation of protective stable transfer films on the counterface surface.
3. Reduction of surface roughness and material removal from the nanocomposite due to the similar size of the fillers and the segments of the polymer chains.

The wear reduction due to transfer films is generally attributed to tribochemical processes and interface reactions, which increase the adhesion of the film to the counterface. It has also been suggested that the nanophases can fill the surface asperities and blend with wear debris, reducing surface roughness.

One of the potential advantages of nanofillers with respect to microfillers [2] is that the nanophases would be less abrasive. In this way, it would be possible to reduce the surface damage caused by abrasion by reducing the size scale of the additives, from micro- to nano-.

One of the key factors influencing the performance of the final material is the additive concentration [5]. In general, when nanophases are used, an approximate reduction by a factor of 10 with respect to the concentration of the corresponding microphases is observed [6].

Such a low additive content makes it possible to use nanophases to improve the properties not only of neat polymers, but also of fibre-reinforced composites [7]. It has been shown that nanoparticles can penetrate fibre bundles while microparticles are not able to do so.

2 Polymer Matrices

A large number of polymer nanocomposites are based on epoxy resins or on thermoplastics with high thermal resistance [4].

Much effort has also been dedicated to the tribological improvement of thermoplastics with applications where the resistance to friction and/or wear are critical.

Politetrafluoroethylene (PTFE) is a thermoplastic widely used as solid lubricant due to the low friction coefficients. However, it has the important drawback of its very high wear rates. There have been many attempts to improve the wear resistance of PTFE by addition of different types of nanoparticles and nanophases, including carbon nanotubes.

Polyamide (PA) is a good bearing material because of its high strength and good wear resistance, so many studies on the friction and wear of PA composites have been reported and its composites have been successfully used in many fields [8]. PA has superior wear resistance sliding against a steel counter-face relative to other polymers. The main factor that influences the friction and wear characteristics of polymer composites are the particle size, morphology, and concentration of the filler. If the particles are large and hard, they are readily pulled out of the matrix material and contribute to wear of the composites by their abrasive action and cause wear and damage of the counterpart material [8].

Ultra-high molecular weight polyethylene (UHMWPE) can be used as a bearing surface and has found applications in hip replacement surgery. However, failures due to wear limit their service life. When excessive wear of UHMWPE occurs, it causes the replacement of the implant or prosthesis. This has lead to an increasing interest in reducing the wear rate of UHMWPE and increase the service life of the components. Nano-hydroxyapatite (nano-HA) has been used [9] as bone repair material due to its excellent bioactive and biocompatibility properties. Nano-HA and PVA have been used to obtain nano-HA/PVA gel composites, which not only improve the mechanical properties of the composites but also its excellent bioactivities.

Polyacrylonitrile (PAN) and polystyrene (PS) thermoplastic are widely used in instructural applications, automotive components, aerospace industry and railway transport. Polystyrene–polyacrylonitrile (PS-SAN) copolymer could show enhanced mechanical properties comparable to the pure PAN and PS [10]. The copolymer can effectively improve the monomer characteristics including heat-resistant, wear resistance, anti-burning and low temperature resistance, machining performance, etc. In addition, SAN is widely used in home wiring, automobile making, commodity and other productions. However, its applications are limited by its relatively weak tribological properties.

3 Reinforcements

3.1 Nanoparticles

Nanoparticles have high specific surface areas and high surface energy due to their small scales [11]. They can react with macromolecular chains chemically or physically to enhance the interactions between the macromolecular chains after they are added to polymer. Thus the friction and wear properties of nanocomposites can be significantly enhanced.

The incorporation of nanosize ceramic particles [4, 6, 8, 11–13] of oxides, carbides or nitrides like TiO_2, SiO_2, ZrO, SiC, Si_3N_4 and Al_2O_3 to polymer matrices has lead to better enhancement in wear resistance. For a detailed review of the effect of ceramic nanoparticles on the friction and wear decrease of polymers see the review by Carrión et al. [4]. One of the more widely used filler are alumina nanoparticles [4, 6, 11].

In the case of polyimide (PI)-based nanocomposites, a content of 3.0–4.0 wt % Al_2O_3 nanoparticles gives the lowest friction coefficient and wear volume. This was attributed to differences in transfer film, worn surface morphologies, and wear debris.

The wear-resistance of PTFE can be increased by 3000 $\times$ with the addition of 1 wt % Al_2O_3 nanoparticles, although the lowest wear rate and friction coefficient was achieved with a 5 wt % concentration.

The addition of alumina nanoparticles has also been found to increase the wear-resistance of other thermoplastics such as polyethyleneterephtalate (PET). The nanocomposites formed a protective transfer layer. The existence of optimum filler content has been explained by the formation of abrasive agglomerates within the transfer films for the higher nanoparticles contents.

The wear rate of polyphenylene sulphide (PPS) was reduced when filled with 1–2 vol. % Al_2O_3 nanoparticles. In this case, the reduction in wear rate was related to the increase in bond strength between the transfer film and the counterface.

The influence of nano-Al_2O_3 on tribological properties of polyoxymethylene (POM) nanocomposites has also been studied [11] under dry and oil lubricated sliding conditions.

The effects of micro-TiO_2 (44 μm) and nano-TiO_2 (10 nm) particles on the wear resistance of epoxy resin have been compared [13].

TiO_2 nanoparticles can remarkably reduce the wear rate of epoxy, while the micro-TiO_2 particles cannot. Similar conclusions were reached in earlier works using micro- and nano-copper particles added to polyoxymethylene (POM), and for micro- and nano-SiC added to polyetheretherketone (PEEK). The results have been explained by a predominantly plastic deformation mechanism in the case of nano-Cu nanocomposites and by the formation of a uniform and stable transfer film in the case of nano-SiC.

Self-lubricating composites containing both nanoparticles and fibre fillers, like nano-$CaCO_3$ and glass fibres, nano-Al_2O_3 and carbon fibres or nano-TiO_2 and short carbon fibres, have been manufactured. These materials can combine the positive effects of both types of reinforcements, being able to operate under more severe sliding conditions than the composites with nanoparticles alone. It was considered that the nanoparticles might have enhanced the adhesion between the reinforcing fibres and the matrix.

3.2 Nanoclays

There exists a great interest in the development of new polymer-layered silicate nanocomposites [14–16], in the expectation of improved physicochemical and mechanical properties with respect to the pure polymers and conventional composites, with the use of a relatively low filler proportion.

Polymer nanocomposites based on layered silicates (e.g., montmorillonite, MMT) have been widely investigated. The montmorillonites are layered silicates which consist of stacked platelets with a thickness of the order of 1 nm. Montmorillonites are naturally hydrophilic and they are modified by ion exchange, e.g., with quaternary ammonium salts, in order to increase their organophilic characteristics. These modified clays, also called organoclays present better compatibility with polymer matrices.

The discovery that the presence of clay nanoparticles can improve the properties of polymers has lead to considerable research efforts focused on polymer nanoclay nanocomposites derived from different polymers such as epoxy, polyimide, polyurethane, polyesteramide, but relatively few studies have been carried out on the tribological properties of polymer nanoclay nanocomposites. The tribological behavior of thermoset-matrix nanocomposites modified by different proportions of organically modified nanoclay has been reported [14]. Carrion et al., prepared and described the friction and wear behaviour of new polycarbonate-organoclay nanocomposites [15].

Some studies have reported that nanoclay improved the tribological properties of other polymers like polyester, polyamide and polyvinylidenefluoride (PVDF) [16].

3.3 Carbon Nanotubes

Carbon nanotubes (CNTs) are becoming one of the most relevant materials in nanoscience and nanotechnology. Their exceptional properties are opening new fields in science and engineering [17].

CNTs are exceptionally stiff and strong. The Young's modulus of single-walled carbon nanotube (SWNT) is estimated to be up to 5 TPa, while the average value of Young's modulus of isolated multi-walled carbon nanotube (MWNT) has been determined to be 1.8 TPa. The tensile strength of CNTs is 100 times stronger than that of steel, while their density is only one-sixth to one-seventh that of steel.

Carbon nanotube nanocomposites have been developed by numerous research groups aiming to produce new novel strong and light composite materials, but they can also be used to prepare new nanocomposites with excellent tribological properties.

Because CNTs can withstand repeated bending, buckling and twisting to build excellent nanocomposite matrices, they can show lower friction coefficient and wear rate compared with the pure substrate matrix [10].

In a relatively recent review, Dasari et al. [18], concluded that new contributions to the knowledge of the tribological behaviour of polymer-carbon nanotubes nanocomposites are needed.

The chemical bond structure of the nanotubes, with a surface free of dangling atoms, makes them good solid lubricants with a friction coefficient against diamond as low as 0.01 [19]. However, the tribological performance of polymers reinforced by carbon nanotubes is not always enhanced.

In this section, some relevant results on the tribological performance of carbon nanotube-reinforced polymers, both with thermoset and thermoplastic matrices, are reviewed.

Some researchers have focused on the improvement of the friction and wear resistance of commodity thermoplastics with relatively poor tribological performance such as PS and SAN [20–22].

Wang et al. [10] have described the development and tribological behaviour of nanocomposites with SAN reinforced by single-walled carbon nanotubes (SWCNT) obtained by in situ polymerization.

Figures 3, 4, 5 show the effect of increasing SWCNT concentration on the hardness (Fig. 1), friction coefficient (Fig. 2) and wear rate (Fig. 3) of SAN-SWCNT nanocomposites. As can be observed, in this case, there is an optimum SWCNT concentration of 1 wt %.

Zarrin et al. [22] added SWCNTs to a polyimide (PI) matrix, to find a relationship between the friction coefficients and the concentration of the additive.

At low concentrations, the properties of the nanocomposite are more related to the elastic behaviour of the carbon nanotubes, while for high concentrations, the nanocomposites properties depend on the plastic behaviour of the carbon nanotubes.

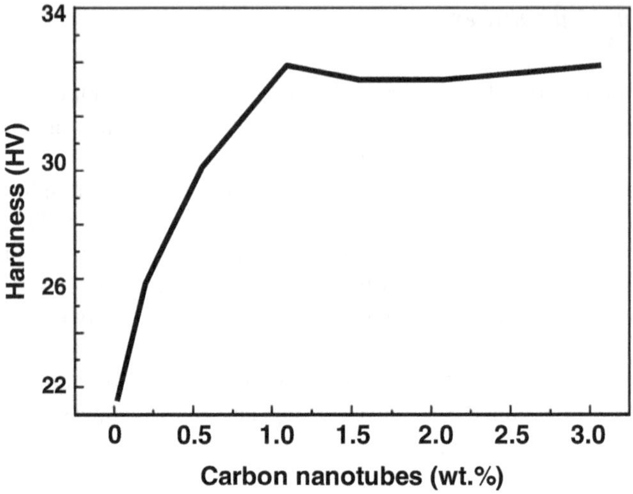

Fig. 3 Hardness of SAN-SWCNT (adapted from reference [10])

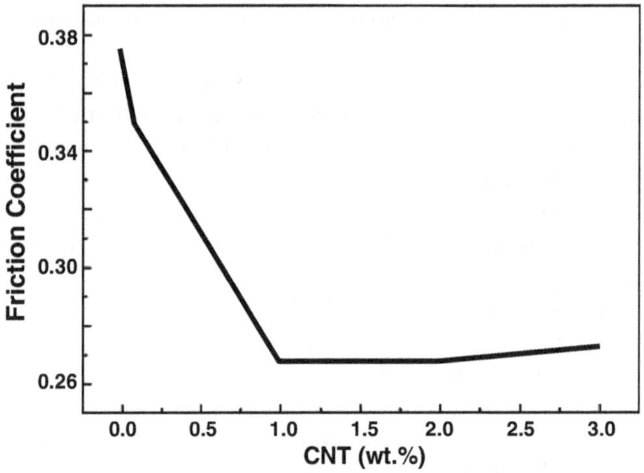

Fig. 4 Friction coefficients of SAN-SWCNT (adapted from reference [10])

Vail et al. [23] studied the influence of SWCNTs additions, between 2 and 15 wt %, on the properties of PTFE, finding a 20 × increase in the wear resistance for a 5 wt % concentration of SWCNTs. However, no direct relationship exists between wear and friction, as for the same concentration of nanotubes of 5 wt %, a 50 % increase in the friction coefficient was observed.

Multi-walled carbon nanotubes (MWCNTs) have also been used to obtain new nanocomposites wih enhanced tribological performance.

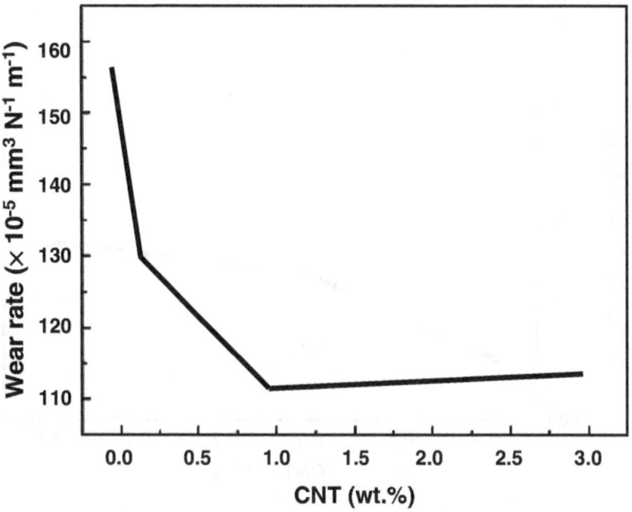

Fig. 5 Wear rates of SAN-SWCNT (adapted from reference [10])

Variable concentration addition of MWCNTs to PTFE [24] yielded the highest mechanical resistance and the lowest wear rates for a 15–20 vol. % of MWCNTs. This effect was attributed to the reduction of both adhesive and abrasive wear. MWCNTs are detached from the polymer matrix during sliding and transferred to the interface, thus preventing the direct contact between the sliding surfaces.

Also using MWCNTs, May et al. [25] have reduced the friction coefficients of polyamides (PA). Using MWCNTs. Meng et al. [26] were able to reduce friction coefficients and wear rates of polyamide 6 (PA6), not only under dry wear conditions, but also under water lubricated sliding.

Giraldo et al. [27, 28], studied the resistance to abrasive wear under scratching of PA6 nanocomposites with different types of carbon nanotubes. The nanocomposites showed lower instantaneous penetration values, but the permanent surface damage, measured as the residual penetration depth after viscoelastic recovery, was not lower for the nanocomposites than for the base polymer.

Other authors [29] have shown that a 0.1 wt % addition of CNTs can be enough to reduce friction and wear of polycaproamides, provided that an homogeneous distribution of the additive is achieved. In this case, the reinforced effect was attributed to the nucleating action of the nanotubes to give a fine spherulitic structure.

Functionalized MWCNTs have also been used to improve the wear resistance of polymethylmethacrylate (PMMA) [30].

Previous studies carried out by our research group [31], have shown the improvement in the abrasive wear resistance of PMMA by addition of SWCNTs.

Several research groups have addressed the increase of the wear life of polymers widely used in tribological applications, such as high density (HDPE) or

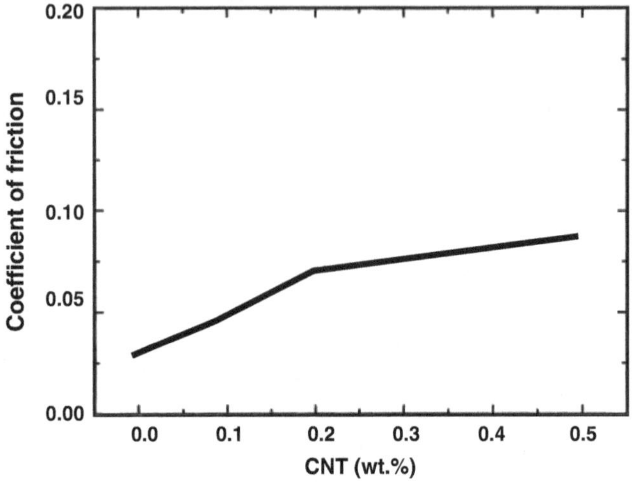

Fig. 6 Evolution of friction coefficients of UHMWPE with CNT content (adapted from reference [32])

ultrahigh density (UHMWPE) polyethylene by the addition of carbon nanotubes [32–35].

The graphs (Figs. 6 and 7) show results described by Zoo et al. [32] on the the influence of an increasing concentration of CNTs on the properties of UHMWPE. A wear reduction with increasing CNT content can be observed (Fig. 5). However, this wear reduction is not accompanied by a friction decrease. In fact, a slight increase with increasing CNT content takes place (Fig. 4).

Both in Figs. 6 and 7, a change in slope occurs for a CNT content of 0.2 wt %. From this concentration to the maximum 0.5 wt % tested, the friction increase and the wear reduction are less pronounced.

MWCNTs were incorporated into a polycarbonate (PC) matrix by Liu et al. [36]. An optimum concentration between 1 and 3 wt % of MWCNTs was found, and the wear rate reduction for the nanocomposite with respect to neat PC was related to the enhancement of the mechanical properties due to the π–π interactions between the aromatic groups of the polymer chains and the carbon nanotubes.

MWCNTs are also able to increase the wear resistance of epoxy resin [37, 38] and of high resistant composites such as those derived from polyetheretherketone (PEEK) with carbon fibre reinforcements [39].

The tribological behaviour of the nanocomposites is a function of, at least, the following factors [38]:

1. The nature of the nanotubes. The manufacturing process and the subsequent purification methods can be determinant.
2. The dispersion method [40].

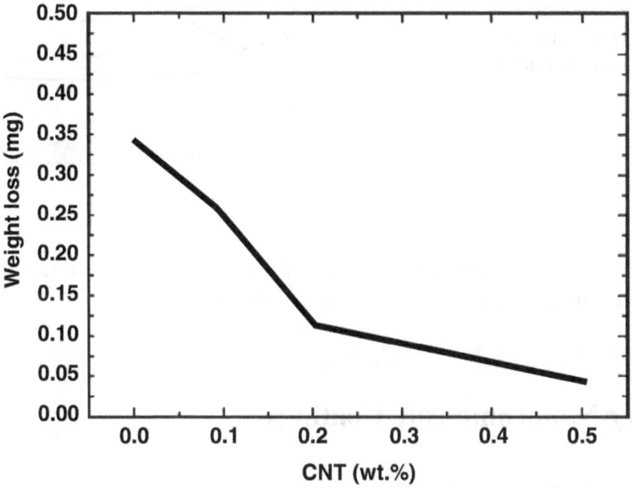

Fig. 7 Evolution of the weight loss due to wear of UHMWPE, with CNT content (adapted from reference [32])

3. The concentration of nanotubes present in the nanocomposites.
4. Chemical treatment and the addition of compatibilizers to improve the interface interaction between the nanotubes and the matrix [17].

Thus, the microhardness of PS nanocomposites containing chemically treated SWCNTs increase sharply when the SWCNTs content is below 1.5 wt %, but decreases when the content is above 1.5 wt %. This is attributed to the agglomeration of SWCNTs in the polymeric matrix. These results suggest that compatibility between carbon nanotubes and polymers is improved with the chemical treatment on the nanotubes.

In addition, if the polymer is formed by in situ polymerization, the mechanical properties and the tribological performance are improved, in some cases, due to a more homogeneous dispersion.

Not only pure polymers have been used to prepare nanocomposites, since also polymer blends have been reported. The wear resistance of the UHMWPE/HDPE composites can be significantly improved by adding MWCNTs. The wear rate decreases with increasing MWCNTs content from 0 to 2 wt % [18].

Carbon nanotubes polymer nanocomposites produced by solution and casting with different concentration on each side of the polymer matrix have been described [17]. Thus, the scratch resistance is different for each side of the nanocomposites, depending on CNT concentration (Fig. 8).

This anisotropic behaviour is attributed to the presence of a gradient distribution in carbon nanotube polymer nanocomposite. Although less carbon nanotubes are found in the upper side (Fig. 8), there is a more uniform distribution in the polymer matrix and therefore load transfer is achieved more effectively. The bottom

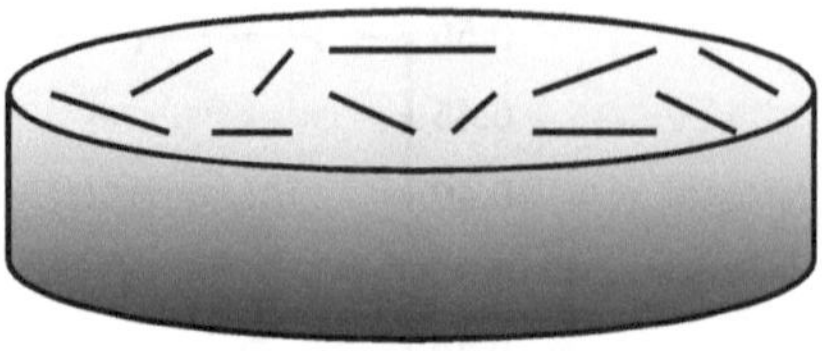

Fig. 8 Carbon nanotube polymer nanocomposite with different concentration in *top* and *bottom* sides [17]

side contains a high concentration of nanotubes, which agglomerate and produce a poor distribution in different regions of the nanocomposite.

4 Polymer Nanocomposite Coatings

For many contacting surfaces in mechanical systems protective coatings are applied on the sliding surfaces.

Among the most relevant properties of a coating for machine element applications are high wear resistance, low friction coefficient and good adhesion to the substrate. If the coated component slides against an uncoated surface, a low wear rate of the counterface is also needed.

The anti-wear coatings most commonly used in industry are based on metal carbides, sulfides, nitrides and oxides. These coatings increase wear resistance but may present some serious disadvantages including high friction, low adhesion and system failure caused by the wear debris from the hard coating materials.

Polymer coatings, with their ability to be coated using simple techniques and their cost effectiveness, present a very viable alternative protecting technology. Polymer thin coatings have shown excellent tribological properties when deposited onto various metallic substrates such as steel and aluminium. The application of a polymer coating reduces the lubricant consumption and increase the life of tribological systems [8].

The mechanical and tribological performances of polycarbonate film and nano-SiO_2/polycarbonate composite coating were studied with micro/nano-scale indentation and scratch tests [41]. The experimental results showed that the hardness and stiffness are increased after the addition of nano-SiO_2. The scratch tests results indicated that the nano-SiO_2/polycarbonate coating exhibits smaller scratch depth and lower frictional coefficient.

The addition of the low inclusion of ZnO nanoparticles and ZnO whiskers can improve the friction reducing and anti-wear abilities of polyurethane (PU) coatings with ZnO whiskers [42], as the filler was superior to ZnO nanoparticles in terms of the ability to decrease friction coefficient and wear rate of the PU coating. The investigations of the worn surfaces showed that the fillers of ZnO nanoparticles and ZnO whiskers were able to enhance the adhesion of the transfer films of the PU coating to the surface of the counterpart, so they significantly reduced the wear rate of the PU coating.

Bautista et al. [43] have shown that nanosized inorganic components improve scratch and wear resistance maintaining low viscosity, gloss and transparency of the nanocomposite coatings.

Tribological and corrosion behaviors of nanostructured tungsten carbide (WC) particles/polymer composite coatings were studied by using microscratch technology and electrochemical technique [44]. The coatings containing nanostructured WC particles showed a significant increase in hardness and scratch resistance compared to that of pure polymer coating. The improvement in hardness and scratch resistance is attributed to the dispersion hardening of nanostructured WC particles in polymer coatings. Corrosion test results showed that the nanostructured WC particles/polymer composite coatings exhibit better or at least equivalent corrosion resistance to that of the pure polymer coating.

The dry friction and wear behaviours of polyamide-1010 (PA-1010) and PA-1010/nanoSiO_2 composite coatings were investigated under dry sliding conditions at room environment [8]. The results showed that the addition of nanometer-sized silica could increase the crystallinity of the coatings, improve the mechanical properties of coatings and effectively reduce friction and wear of pure PA-1010. The wear mechanism of the coatings was explained in terms of plastic deformation, fatigue tearing and adhesive abrasion. In addition, the nanocomposite coating containing 1.5 wt % nanometer-sized silica displayed better properties.

The tribological and electrochemical corrosion properties of Al_2O_3/polymer nanocomposite coatings were studied [45] by using micro-hardness test, single-pass scratch test and abrasive wear test. The coatings containing Al_2O_3 nanoparticles showed improvement in scratch and abrasive resistance compared with that of the neat polymer coating. The improvement in scratch and abrasive resistance is attributed to the dispersion hardening of Al_2O_3 nanoparticles in polymer coatings. Corrosion test results showed that the embedded Al_2O_3 nanoparticles in polymer matrix do not decrease the corrosion protection ability of the polymer.

4.1 Polymer Nanocomposite Coatings with Carbon Nanotubes

A nanocomposite coating of UHMWPE reinforced with 0.1 wt % single-walled carbon nanotubes (SWCNTs), was developed [46] and its tribological properties studied against three different counterface materials, silicon nitride, AISI 52100 bearing steel and brass. It was found that the tribological properties of the nanocomposite coating were superior and did not alter even after being exposed to UV radiation.

Although CNTs have these excellent properties, they are not easily dispersed and tend to form large agglomerates, which influence the mechanical properties and tribological behavior of polymeric composites.

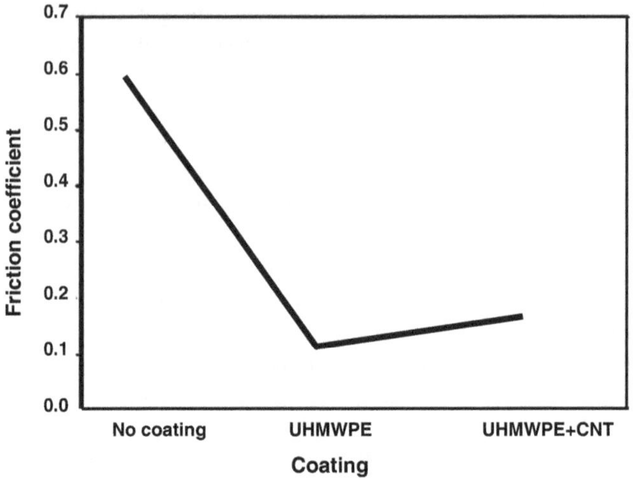

Fig. 9 Comparison of the coefficients of friction and wear life for the uncoated Al shaft, UHMWPE coated and UHMWPE + SWCNTs coated shafts against Al flat plate under dry sliding conditions (adapted from reference [48])

Many studies have investigated different techniques for improving CNT dispersion in polymer matrices by introducing chemical functionalization to CNT surfaces. The use of chemically modified CNTs may help eliminate or diminish this inconvenient by the attachment of appropriate chemical moieties to the surfaces of CNTs, through either van der Waals interactions or covalent bonds.

The tribological behavior of PMMA-grafted-MWCNT nanocomposite coatings prepared by applying the direct casting method on aluminium surfaces was investigated [47]. Amino-functionalized MWCNTs were first covalently functionalized with PMMA to introduce the interaction force. The PMMA-grafted-MWCNT exhibited a higher dispersive capacity in a polyimide matrix than pure MWCNT composite due to strong interfacial interactions in the nanocomposite. The maximum possible amount of PMMA-grafted-MWCNT was 5 wt % before fragilization takes place.

The nanocomposite coating reinforced with 3 wt % MWCNT was the most stable and had the lowest coefficient of friction in this study. This was attributed to the formation of a thin, uniform friction film on the wear track during sliding, as the friction film plays an important role in facilitating the stable frictional behavior of the nanocomposite coatings.

The lowest coefficient of friction was obtained for the nanocomposite coating reinforced with 3 wt % MWCNT-grafted to PMMA. A uniform, thin friction film on the counterface was formed in this case. Meanwhile, rupture and fragmentation of the other coatings were observed, and were attributed to higher friction coefficients and unstable frictional behaviors of the coatings during sliding. The conclusion was that the uniform distribution of MWCNTs in the matrix is a prerequisite for lowering friction, and the amount of MWCNTs used for coating should not be excessive in order to avoid brittleness.

A nanocomposite polymer coating of UHMWPE reinforced with 0.1 wt % of single-walled carbon nanotubes (SWCNTs) has been developed [48] and coated on steel substrates for possible applications as a boundary lubricant in bearings and gears (Fig. 9).

The addition of SWCNTs to the polymer matrix not only helps improving the mechanical properties, such as hardness and the load bearing capacity of the coating, but also enhances its frictional and wear properties at elevated temperatures.

The boundary lubrication regime plays a very important role in determining the life span of any of the two mating parts under liquid-lubricated conditions. It is during the start\stop cycles when insufficient fluid is available to fully separate the surfaces in relative motion and thus unusually high wear takes place; a case of boundary lubrication.

Samad et al. [49], have studied the feasibility of using polymer coatings as boundary lubricants, using UHMWPE films coated on aluminium substrates under dry and lubricated conditions. In order to increase the load bearing capacity of the UHMWPE coatings, 0.1 wt % of single-walled carbon nanotubes were added. Stribeck curves have been generated to evaluate the effectiveness of the pristine UHMWPE and the nanocomposite coatings in the various regimes of lubrication, especially the boundary lubrication regime. It is observed that the selected polymer coatings are effective in protecting the metallic surfaces without causing any observable oil contamination with wear debris.

Under boundary-lubrication conditions, the nanocomposite polymer film was found to be very effective in protecting the mating surfaces from wear under the base oil lubrication without any additives.

The addition of 0.1 wt % CNTs to the UHMWPE polymer improved the wear life of the coating from 250,000 cycles to more than 2 million cycles under dry sliding conditions against an aluminium plate.

Minimum wear of the coating and no wear of the counterface metal were observed after 2 million cycles of sliding in lubricated contact with the nanocomposite film.

5 Concluding Remarks

In the present review, we have shown some aspects of the current state of development of polymer nanocomposites with enhanced tribological performance. Some of the latest developments are centred on the improvement of the homogeneity of the dispersions by functionalization, chemical modification, heat treatment or the addition of compatibilizers between the polymer matrix and the nanophases [50–57].

Despite the spectacular growth experienced by this field of research during the last two decades, significant advances are still needed in order to understand and be able to control the nature, concentration and dispersion of the nanofiller, the interface interactions with the matrix and their relationship with friction and wear.

References

1. Friedrich K, Lu Z, Hager AM (1995) Recent advances in polymer composites tribology. Wear 190:139–144
2. Chang L, Zhang Z, Ye L, Friedrich K (2007) Tribological properties of epoxy nanocomposites III. Characteristics of transfer films. Wear 262:699–706
3. McCook NL, Hamilton MA, Burris DL, Sawyer WG (2007) Tribological results of PEEK nanocomposites in dry sliding against 440C in various gas environments. Wear 262:1511–1515
4. Carrión-Vilches FJ, Bermúdez MD (2008) Nanoparticles in polymer nanocomposites for tribological applications. In: Frisiras CT (ed) Progress in nanoparticles research. Nova Science Publishers Inc. New York, pp 219–244
5. Larsen TØ, Andersen TL, Thorning B, Horsewell A, Vigild ME (2008) Changes in the tribological behavior of an epoxy resin by incorporating CuO nanoparticles and PTFE microparticles. Wear 265:203–213
6. Sawyer GW, Freudenberg KD, Bhimaraj P, Chadler LS (2003) A study on the friction and wear of PTFE filled with alumina nanoparticles. Wear 254:573–580
7. Larsen TØ, Andersen TL, Thorning B, Vigild ME (2008) The effect of particle addition and fibrous reinforcement on epoxy-matrix composites for severe sliding conditions. Wear 264:857–868
8. Li YD, Ma Y, Xie B, Cao S, Wu Z (2007) Dry friction and wear behavior of flame-sprayed polyamide1010/n-SiO_2 composite coatings. Wear 262:1232–1238
9. Pan Y, Xiong D (2009) Friction properties of nano-hydroxyapatite reinforced poly(vinyl alcohol) gel composites as an articular cartilage. Wear 266:699–703
10. Wang C, Xue T, Dong B, Wang Z, Li HL (2008) Polystyrene–acrylonitrile–CNTs nanocomposites preparations and tribological behavior research. Wear 265:1923–1926
11. Sun LH, Yang ZG, Li XH (2008) Study on the friction and wear behaviour of POM/Al_2O_3 nanocomposites. Wear 264:693–700
12. Guo QB, Rong MZ, Jia GL, Lauc KT, Zhang MQ (2009) Sliding wear performance of nano-SiO_2/short carbon fiber/epoxy hybrid composites. Wear 266:658–665
13. Zhang Z, Breidt C, Chang L, Haupert F, Friedrich K (2004) Enhancement of the wear resistance of epoxy: short carbon fibre, graphite, PTFE and nano-TiO_2. Compos A Appl Sci Manuf 35:1385–1392
14. Jawahar P, Gnanamoorthy R, Balasubramanian M (2006) Tribological behaviour of clay—thermoset polyester nanocomposites. Wear 261:835–840
15. Carrión FJ, Arribas A, Bermúdez MD, Guillamon A (2008) Physical and tribological properties of a new polycarbonate-organoclay nanocomposite. Eur Polymer J 44:968–977
16. Peng QY, Cong PH, Liu XJ, Liu TX, Huang S, Li TS (2009) The preparation of PVDF/clay nanocomposites and the investigation of their tribological properties. Wear 266:713–720
17. Martinez-Hernandez AL, Velasco-Santos C, Castaño VM (2010) Carbon nanotubes composites: processing, grafting and mechanical and thermal properties. Current Nanosci 6:12–39
18. Dasari A, Yu ZZ, Mai YW (2009) Fundamental aspects and recent progress on wear/scratch damage in polymer nanocomposites. Mater Sci Eng R 63:31–80
19. Mylvaganam K, Zhang LC, Xiao KQ (2009) Origin of friction in films of horizontally oriented carbon nanotubes sliding against diamond. Carbon 47:1693–1700
20. Yang Z, Dong B, Huang Y, Liu L, Yan FY, Li HL (2005) Enhanced wear resistance and micro-hardness of polystyrene nanocomposites by carbon nanotubes. Mater Chem Phys 94:109–113
21. Liu S, Hsu W, Chang K, Yeh J (2009) Enhancement of the surface and bulk mechanical properties of polystyrene through the incorporation of raw multiwalled nano-tubes with the twin-screw mixing technique. J Appl Polym Sci 113:992–999
22. Zarrin T, Ribeiro R, Banda S, Ounaies Z, Liang H (2009) Effect of SWCNT on tribological behavior of polymeric nanocomposite. In: Proceedings of the Stle/ASME international joint tribology conference pp 103–105

23. Vail JR, Burris DL, Sawyer GW (2009) Multifunctionality of single-walled carbon nanotube-polytetrafluoroethylene nanocomposites. Wear 267:619–624
24. Chen WX, Li F, Han G, Xia JB, Wang LY, Tu JP, Xu ZD (2003) Tribological behavior of carbon-nanotube-filled PTFE composites. Tribol Lett 15:275–278
25. May B, Hartwich MR, Stengler R, Hu XG (2009) The influence of carbon nanotubes on the tribological behavior and wear resistance of a polyamide nanocomposite. Adv Tribol 515–515
26. Meng H, Sui GX, Xie GY, Yang R (2009) Friction and wear behavior of carbon nanotubes reinforced polyamide 6 composites under dry sliding and water lubricated condition. Compos Sci Technol 69:606–611
27. Giraldo LF, Lopez BL, Brostow W (2009) Effect of the type of carbon nanotubes on tribological properties of polyamide 6. Polym Eng Sci 49:896–902
28. Giraldo LF, Brostow W, Devaux E, Lopez BL, Perez LD (2008) Scratch and wear resistance of polyamide 6 reinforced with multiwall carbon nanotubes. J Nanosci Nanotechnol 8:3176–3183
29. Krasnov AP, Afronicheva OV, Mit' VA, Bazhenova VB, Volkova TV, Sinitsyna OV, Vygodskii YS, Rashkovan IA, Kazakov ME (2009) Role of nanofiller in friction of polymers based on polycaproamide: indirect effect. J Friction Wear 30:350–355
30. Huang YL, Yuen SM, Ma CCM, Chuang CY, Yu KC, Teng CC, Tien HW, Chiu YC, Wu SY, Liao SH, Weng FB (2009) Morphological, electrical, electromagnetic interference (EMI) shielding, and tribological properties of functionalized multi-walled carbon nanotube/polymethyl methacrylate (PMMA) composites. Compos Sci Technol 69:1991–1996
31. Arribas A, Bermúdez MD, Carrión FJ, Espejo C, Martínez-López E, Sanes J (2011) Study of the scratch resistance of polymethylmethacrylate-single walled carbon nanotubes nanocomposites. Effect of modification by a room temperature ionic liquid. In: David JP (ed) Materials and surface engineering. Woodhead Publishing, Oxford, 2011, pp 1–22
32. Zoo YS, An JW, Lim DP, Lim DS (2004) Effect of carbon nanotube addition on tribological behavior of UHMWPE. Tribol Lett 16:305–309
33. Kanagaraj S, Varanda FR, Zhil'tsova TV, Oliveira MSA, Simoes JAO (2007) Mechanical properties of high density polyethylene/carbon nanotube composites. Compos Sci Technol 67:3071–3077
34. Wang M, Hsu T, Zheng J (2009) Sintering process and mechanical property of MWCNTs/HDPE bulk composite. Polym Plast Technol Eng 48:821–826
35. Lee J, Kathi J, Rhee KY, Lee JH (2010) Wear properties of 3-Aminopropyltriethoxysilane-functionalized carbon nanotubes reinforced ultra high molecular weight polyethylene nanocomposites. Polym Eng Sci 50:1433–1439
36. Liu S, Hwang S, Yeh J, Pan K (2010) Enhancement of surface and bulk mechanical properties of polycarbonate through the incorporation of raw MWNTs—Using the twin-screw extruder mixed technique. Int Commun Heat Mass Transfer 37:809–814
37. Armstrong G, Ruether M, Blighe F, Blau W (2009) Functionalised multi-walled carbon nanotubes for epoxy nanocomposites with improved performance. Polym Int 58:1002–1009
38. Jacobs O, Xu W, Schadel B, Wu W (2006) Wear behaviour of carbon nanotube reinforced epoxy resin composites. Tribol Lett 23:65–75
39. Li J, Zhang LQ (2009) The effects of adding carbon nanotubes to the mechanical and tribological properties of a carbon fibre reinforced polyether ether ketone composite. Proc Inst Mech Eng Part C J Mech Eng Sci 223:2501–2507
40. Chen H, Jacobs O, Wu W, Ruediger G, Schaedel B (2007) Effect of dispersion method on tribological properties of carbon nanotube reinforced epoxy resin composites. Polym Test 26:351–360
41. Wang ZZ, Gu P, Zhang Z (2010) Indentation and scratch behavior of nano-SiO(2)/polycarbonate composite coating at the micro/nano-scale. Wear 269:21–25
42. Song HJ, Zhang ZZ, Men XH, Luo ZZ (2010) A study of the tribological behavior of nano-ZnO-filled polyurethane composite coatings. Wear 269:79–85
43. Bautista Y, Gonzalez J, Gilabert J, Ibañez MJ, Sanz V (2011) Correlation between the wear resistance, and the scratch resistance, for nanocomposite coatings. Prog Org Coat 70:178–185

44. Wang Y, Lim S (2007) Tribological behavior of nanostructured WC particles/polymer coatings. Wear 262:1097–1101
45. Wang Y, Lim S, Luo JL, Xu ZH (2006) Tribological and corrosion behaviors of Al_2O_3/polymer nanocomposite coatings. Wear 260:976–983
46. Samad MA, Sinha SK (2011) Effects of counterface material and UV radiation on the tribological performance of a UHMWPE/CNT nanocomposite coating on steel substrates. Wear 271:2759–2765
47. Kim J, Im H, Cho MH (2011) Tribological performance of fluorinated polyimide-based nanocomposite coatings reinforced with PMMA-grafted-MWCNT. Wear 271:1029–1038
48. Samad MA, Sinha SK (2011) Dry sliding and boundary lubrication performance of a UHMWPE/CNTs nanocomposite coating on steel substrates at elevated temperatures. Wear 270:395–402
49. Samad MA, Sinha SK (2010) Nanocomposite UHMWPE-CNT polymer coatings for boundary lubrication on aluminium substrates. Tribol Lett 38:301–311
50. Men XH, Zhang ZZ, Song HJ, Wang K, Jiang W (2008) Functionalization of carbon nanotubes to improve the tribological properties of poly (furfurylalcohol) composite coatings. Compos Sci Technol 68:1042–1049
51. Men XH, Zhang ZZ, Yang J, Zhu XT, Wang K, Jiang W (2011) Effect of different functional carbon nanotubes on the tribological behaviors of Poly(Furfuryl Alcohol)-derived carbon nanocomposites. Tribol Trans 54:265–274
52. Pollanen M, Pirinen S, Suvanto M, Pakkanen TT (2011) Influence of carbon nanotube-polymeric compatibilizer masterbatches on morphological, thermal, mechanical, and tribological properties of polyethylene. Compos Sci Technol 71:1353–1360
53. Carrión FJ, Espejo C, Sanes J, Bermudez MD (2010) Single-walled carbon nanotubes modified by ionic liquid as antiwear additives of thermoplastics. Compos Sci Technol 70:2160–2167
54. Bermudez MD, Carrion FJ, Espejo C, Martinez-Lopez E, Sanes J (2011) Abrasive wear under multiscratching of polystyrene plus single-walled carbon nanotube nanocomposites. Effect of sliding direction and modification by ionic liquid. Appl Surf Sci 257:9073–9081
55. Coban O, Bora MO, Avcu E, Sinmazcelik T (2011) The Influence of Annealing on the Crystallization and Tribological Behavior of MWNT/PEEK Nanocomposites. Polym Compos 32:1766–1771
56. Shi YJ, Mu LW, Feng X, Lu XH (2011) Tribological behavior of carbon nanotube and polytetrafluoroethylene filled polyimide composites under different lubricated conditions. J Appl Polym Sci 121:1574–1578
57. Carrion FJ, Sanes J, Bermudez MD, Arribas A (2011) New single-walled carbon nanotubes-ionic liquid lubricant. Application to polycarbonate-stainless steel sliding contact. Tribol Lett 41:199–207

Nano and Micro PTFE for Surface Lubrication of Carbon Fabric Reinforced Polyethersulphone Composites

Jayashree Bijwe and Mohit Sharma

Abstract Carbon fibre/fabric (CF) is a privileged reinforcement for advanced polymer composites in tribological applications where performance is primary criteria rather than the cost due to the combination of distinct properties such as high specific strength, thermo-oxidative resistance, thermal and electrical conductivity along with self-lubricity etc. However problems associated with its chemical inertness and surface lipophobicity towards majority of matrix materials need special attention, as these directly affect final properties of a composite. From tribological point of view, surface engineering of composites is very much advantageous in addition to fibrous reinforcement. This chapter reports on the newly developed technique to modify the surface of carbon fabric—Polyethersulphone (CF-PES) composite with Polytetrafluoroethylene (PTFE) micro and nano particles; to improve the tribological properties. Prior to the composite preparation plasma treatment was employed for CF surface modification to promote fiber matrix interfacial adhesion and mechanical interlocking which further improves final composites strength and wear resistance properties. Both the treatment methods; first for fiber surface alteration and secondly for composite surface, proved beneficial to enhance composites performance. The inclusion of nano scale PTFE particles on the surface of a composite proved to be more efficient than the micro-scale PTFE particles to improve tribo-performance of CF-PES composite.

J. Bijwe (✉)
Industrial Tribology Machine Dynamics and Maintenance Engineering Centre (ITMMEC), Indian Institute of Technology, Delhi, Hauz Khas, New Delhi 110016, India
e-mail: jbijwe@gmail.com

M. Sharma
Leibniz-Institut für Polymerforschung Dresden, e.V. Hohe Strasse 6, 01069 Dresden, Germany
e-mail: mohittiet@gmail.com

J. Paulo Davim (ed.), *Tribology of Nanocomposites*, Materials Forming, Machining and Tribology, DOI: 10.1007/978-3-642-33882-3_2,

1 Introduction

1.1 Fabric Reinforced Polymer Composites

For most applications virgin polymers are not the right choices mainly because of poor strength properties. Reinforcements for polymers, in various forms such as particulates, spheres (hollow, solid etc.) and fibrous (short, long, woven, nonwoven etc.) are preferred depending on prerequisite application. Each reinforcement form has its own advantages and limitations. Short fibers, for example offer easy injection moldability for thermoplastic polymer composites. The strength offered, however, is moderate and depends on fiber alignment with respect to the loading direction. Long fibers on the other hand offer very high strength, but only in one direction and that too at the cost of easy processability. These are generally processed by compression molding and hence fiber handling is a tough job. Fabric reinforcement on the other hand offers very high strength in two directions along with easier processability compared to the composites with long fibers. For tribo—materials, most popular fibrous reinforcements consists of glass, carbon, graphite, Aramid etc. Again each has its own advantages and limitations. Glass fibers are least expensive and offer moderate strength and wear resistance (W_R) at the cost of increased coefficient of friction (μ); damage the counterface, by abrasion generally and are used in combination with solid lubricants (SLs). Aramid fibers are of moderate cost, offer considerable W_R resistance and strength without excessive incremental in the μ value, neither damages the counterface. However, their temperature resistance is poor. On the other hand Carbon/graphite fibers are most expensive with excellent; specific strength, thermal conductivity and self-lubricity properties.

1.2 Solid Lubricants for Improving Tribological Performance

For tribological applications, advanced polymer composites are preferably used with fibers/fabric reinforcements along with SLs. SLs have lower surface energy and offer less resistance to shearing and hence low μ values. For tribological purpose the most popular SLs are PTFE, graphite, white graphite (hexa boron nitride/hBN), MoS_2 etc. [1, 2].

1.3 Need for Surface Tailoring of Composites

While designing the high performance tribo-composite (e.g. dry bearing) which can survive under harsh operating conditions; matrix, fillers and reinforcement are selected very judiciously. Such specialty polymers and reinforcements are

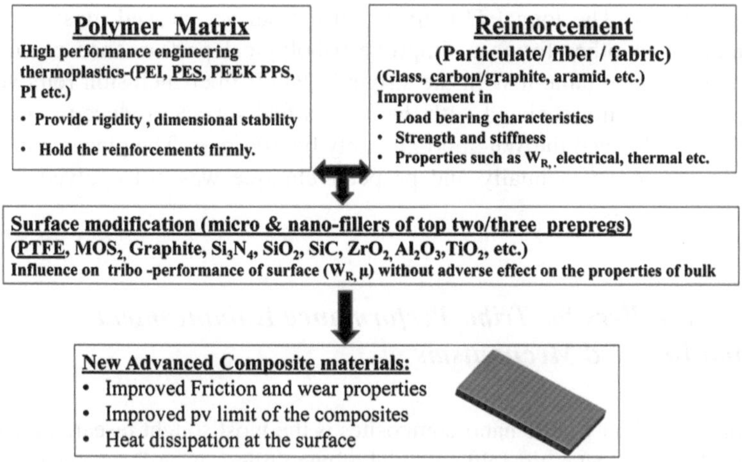

Fig. 1 Schematic surface designing of Polymer Composites

essential but generally expensive. Since tribological composites need high performing surfaces from friction and wear point of view, use of these expensive materials in the bulk is not essential always. SLs being low surface energy materials if added in the bulk of composite, improves tribo-performance at the cost of significant deterioration in the strength apart from unnecessary increase in the cost. It will be wise use the SLs only on surface rather than in the bulk or implement the concept of graded composites in which surface, subsurface and bulk are tailored judiciously with various matrices and reinforcements in such a way that the desired performance can be achieved with adequate cost. (For dry bearings, surfaces should have very low μ, low wear, high thermal conductivity, low expansion and high counter face friendliness, fatigue resistance while bulk should have desired mechanical strength and high thermal conductivity). Interestingly no such efforts are reported in open literature though peripheral information is available in few patent forms [3, 4].

Figure 1 shows the schematic to signpost the judicious importance of each constituent for surface designed advanced polymeric composites which finally attributes to enhanced tribological performance.

A little is reported on the exploration of concept of surface engineering of polymeric bearings [5]. Bijwe et al. [5] prepared surface-tailored composites based on commingled yarns of carbon fiber and Polyetheretherketone (PEEK) using autoclave method. The composites were surface modified with micro sized particles of graphite, MoS_2, copper and PTFE either in isolation or in combination in different proportions in the top fabric layers rather than their inclusion in the bulk. PTFE in various forms, such as particulate (micron sized), wool, short fibers, long fibers, etc., was used to investigate benefits endowed by the surface modifications. Among all SLs PTFE proved most promising. The long PTFE fibers on the surface proved most beneficial as compared to other forms to improve tribological performance of composites without appreciable loss in the strength. PTFE fiber inclusion removed the stick–slip problem associated with the unmodified surface; reduced μ from 0.6 to 0.12 and enhanced the W_R approximately by 70 times. The placement of SLs however, was done manually and proper technique was not evolved in these preliminary studies.

1.4 Nano-Fillers for Tribo-Performance Enhancement and Involved Mechanisms

Development of polymeric nano-composites is the most sought research area from last decade due to the multi-fold potential of nano fillers as performance enhancers when added in small doses [5–13]. The prominent features of nano-fillers are;

1. Nano-particles (NPs) have a very high surface area to volume ratio and hence provide very large interfacial surface area, as a driving force for enhanced interaction with other surface, diffusion, especially at elevated temperature etc.
2. A very low content (generally <2–3 wt %) provides exceptional increase in mechanical strength properties apart from thermal, electrical and biological.

During the wear process, NPs are removed from the surface of a matrix and can act as a third body element in the contacting regions. The rolling effect of the NPs, especially at the edge of exposed fibers, reduce the shear stress in the contact region and hence the μ [14]. This leads to the spontaneous reduction of grooving/cutting wear by the hard counterpart asperities and smoothening of topography of a surface of a composite. It also protects the fibers adhering to the matrix and results in increased fiber thinning rather than breaking before final removal of fibers from the matrix [6–8, 15]. The rolling effect of NPs attributes to the reduction in μ and hence the frictional heating at the tribo couple. The rolling effect is also observed in the case of micro particulate inclusions, for which the small particles tend to tumble through contact region and larger particles plough through it [5, 15]. There is a critical value of the size of a particle governing their transition from rolling to ploughing. To achieve the rolling, the ratio of maximum particle dimension to the minimum gap of contacting bodies must exceed the critical value which depends on the particle itself [16]. The hard micro sized particles and fillers may abrade the counterface. This prevents the formation of a protective transfer film, which increase the counter face roughness and hence the μ of the composite [17]. The NPs have the potential to reduce the abrasion that leads to these cascading and problematic events NPs (<100 nm) are of the same size as the counterface asperities and polish the tallest asperities and promote the development of transfer films. The transfer films shield the composite from direct asperity contact and damage [14]. This film converts the adhesive wear to "like on like" sliding pairs and hence the severity of wear reduces drastically [7, 18].

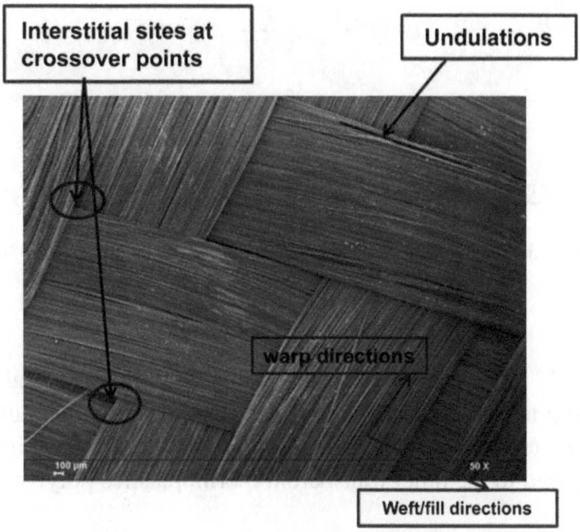

Fig. 2 Fiber architecture of 2 × 2 twill weave carbon fabric [20]

The quantification of advantages in tribology due to variation in size of fillers (nano-sub-micro, micro etc.) is essential. Unfortunately it is not addressed in the literature and needs to be investigated in details. Recently authors have reported on such efforts using PTFE as a solid lubricant for surface modification, Polyethersulphone (PES) as a matrix and carbon fabric (surface treated and untreated) as a reinforcement [19] and the essence of the findings are reported in the subsequent sections.

2 Materials and Methodology

2.1 Details of Selected Materials and Methods

2.1.1 Reinforcement

The carbon fabric (CF) 3 K, 2 × 2 twill weave (Fig. 2) was used as reinforcement and was procured from Fiber Glast Corporation, USA.

2.1.2 Matrix Material

Thermoplastic Polyethersulphone (PES) Veradel 3600P procured from Solvay Advanced Polymer India was selected as a matrix material for development of composites. PES is an amorphous, amber colored, transparent, high performance, heat-resistant and semi tough engineering thermoplastic polymer having density

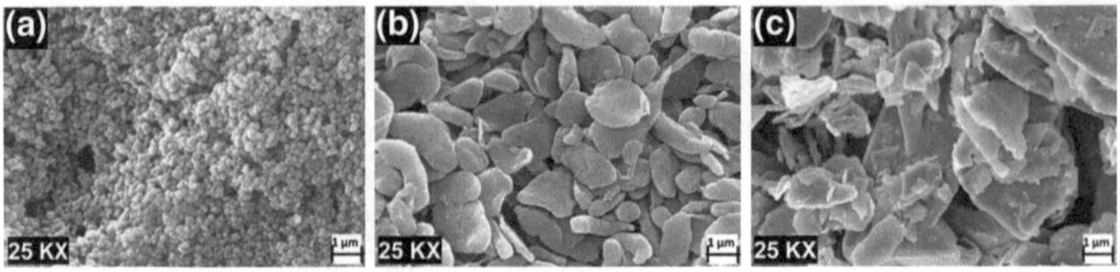

Fig. 3 FESEM micrographs of selected PTFE particles [19, 20]

1.37 g/cm^3. It has a good thermal stability and high continuous use temperature (up to 200 °C). PES has a glass transition temperature (T_g) of 215 °C and a melting temperature (T_m) range of 300–380 °C. It has high hydrolytic stability as compared to the other transparent thermoplastics polymers.

2.1.3 Selection of a Solid Lubricant: Polytetrafluoroethylene

PTFE is a white colored thermoplastic crystalline polymer with a density of 2.2 g/cm^3. Its T_g and T_m are −20 and 321 °C; respectively. Due to the robust nature of molecular bonds in its structure; PTFE is highly resistive to UV radiation and most of the chemicals except alkali metals and elemental fluorine. It retains these properties over a very wide range of temperatures. For surface modification of composites, three sizes of PTFE (micro- 400–450 nm, sub-micro- 200–250 nm and nano- 50–80 nm) as confirmed from FESEM studies (Fig. 3) were selected.

2.2 Surface Treatment of Carbon Fabric

2.2.1 Cold Remote Nitrogen Oxygen Plasma

Generally fibers are not always compatible with the polymer matrices and are provided with various types of sizing by the supplier to enhance their wettability with the selected matrices. Carbon fibers are known for their inertness towards the matrices leading to less performing composite and hence require special attention after procuring from the supplier. Various CF surface treatments; such as acid, plasma, rare earth, gamma treatment are reported successfully in the literature with varying benefits [21–24] and newer and more effective methods are continuously being tried by researchers.

Classical plasma treatment is a well-proven technique for improving adhesion between fiber and matrix. Its improvisation for enhancing effectiveness is one of the most sought research areas. In this regards, cold remote nitrogen plasma (CRNP) treatment with 0.5 % oxygen is a modified process and recently proved

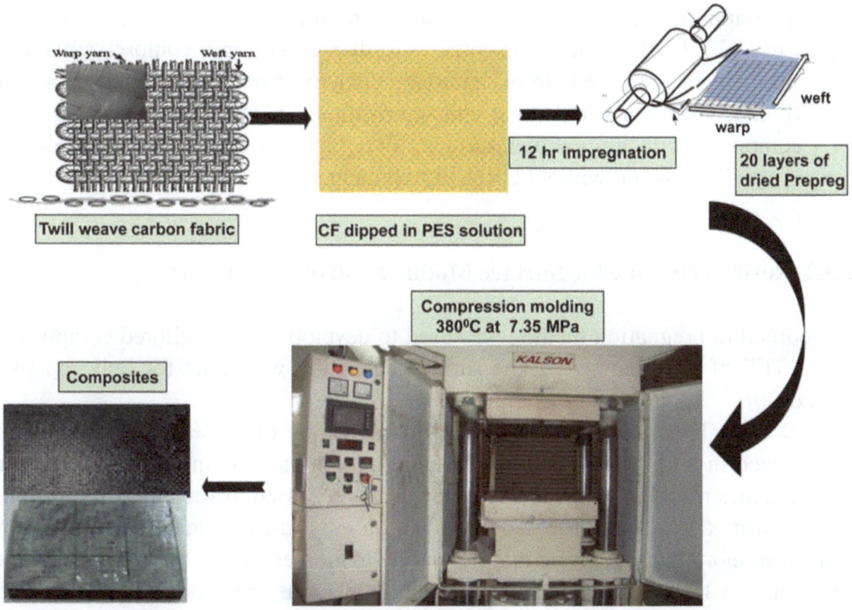

Fig. 4 Schematic of fabrication of CF-PES composites using compression molding [20]

to be successful [22]. The carbon fabric modified with this Cold Remote Nitrogen Oxygen Plasma (CRNOP) technique this was used for reinforcement. Unmodified CF was also selected to quantify the benefits due to surface modification of fabric. (The work involves two surface modifications; first that of fibers and second that of surface of a composite with PTFE.)

2.3 Development of Composites

2.3.1 Selection of a Processing Technique

Solution impregnation technique was selected since it leads to homogenous distribution of a matrix throughout the prepregs including cross-over points in the weave which results the best performance of the composites. Figure 4 shows the schematic for fabrication of composites [20]. Twenty pieces of CF plies (28 × 28 cm) were cut from the roll were immersed in the solution of PES in dichloromethane (DCM) (20 wt %) for twelve hrs in a properly sealed steel container. The prepregs were taken out carefully avoiding the misalignment of the weave and dried in an oven for an hour at 100 °C in a stretched condition and were stacked in a steel mold. PTFE coated glass fabric was placed on the top and bottom of the stacked prepregs as a mold release agent. The mold was then heated in a compression molding machine to a temperature of 380–390 °C for 20 min under a pressure of 7.3 MPa. The composites were then cooled in a compressed condition and then cut with the help of diamond cutter for different mechanical (as per ASTM standards) and tribological characterizations. This composite was treated as a composite with unmodified surface. Two such composites were developed containing CF with and without CRNOP treatment.

2.3.2 Novel Technique for Surface Modification of Composites

A modified impregnation method was used to develop surface tailored composites with PTFE of different sizes. The surface designing was done for only top two layers.

PES and PTFE powders in a selected composition (2 wt % of PTFE in PES) were mixed in a high shear ball mill using zirconia balls in an alcohol media for 16 h. Batches prepared with all PTFE powders were dried in vacuum oven for 2 h. The dried mix was then probe sonicated in an ethyl alcohol medium for 20 min to achieve more homogeneous mixing and de-agglomeration of NPs. The solution impregnation technique (discussed in Sect. 2.3.1) was then used to prepare two prepregs for the surface.

The sequence of applying temperature and pressure was optimized (temperature from 280 to 380 °C in the steps of 20 °C and pressure in the steps of 1 MPa up to 6 MPa). At higher applied temperature and pressure, matrix bleeding and displacement of fibers were observed. Hence, the two tailored prepregs with eighteen untailored prepregs were compression molded under optimized conditions followed by natural cooling under compressed state to the ambient temperature.

3 Characterization of Carbon Fibers

Various surface characterization techniques were adopted to analyze the effect of plasma treatment on the CF surface.

3.1 Field Emission Scanning Electron Microscopy

The surfaces of fibers prior and after treatment were examined by Field emission Scanning Electron Microscopy (FESEM). Figure 5 shows the FESEM images of untreated and plasma treated fibers indicating the increased perforations and roughness of the fiber due to the treatment, which was responsible for better fiber-matrix mechanical interlocking and hence enhanced adhesion. High resolution FESEM micrographs in Fig. 5b indicate deeper and narrower ridges along the length of the treated fibers. In the process of composite manufacturing during

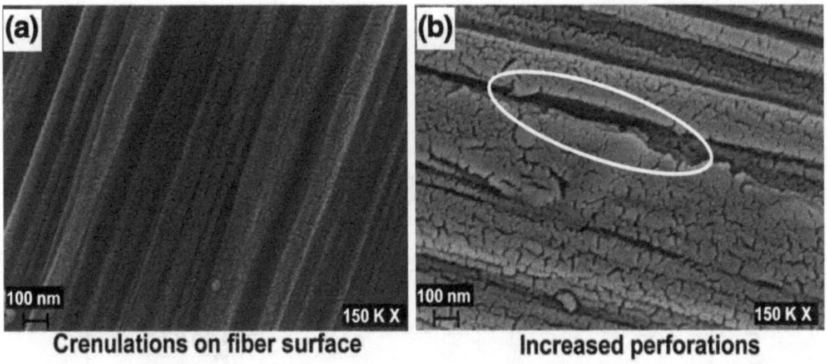

Fig. 5 *High* resolution FESEM (Mag. ×150 K) micrographs of a carbon fiber **a** before treatment **b** after plasma treatment [19, 20]

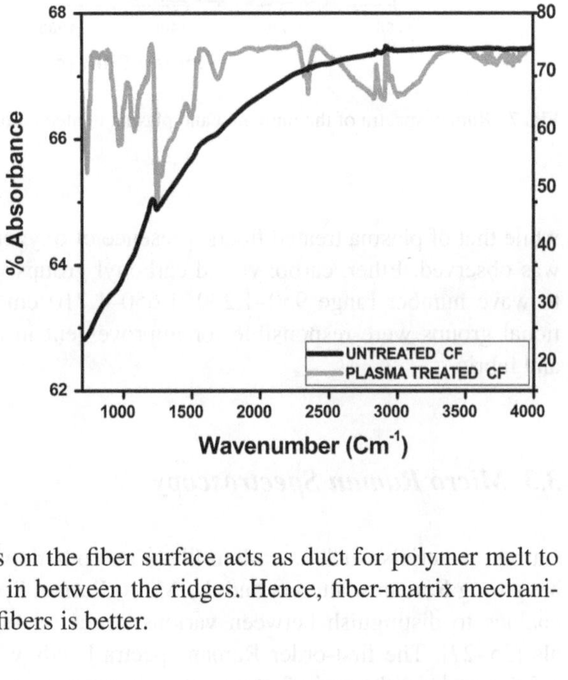

Fig. 6 ATR-FTIR spectra of untreated and plasma treated CF [22]

compression molding, groves on the fiber surface acts as duct for polymer melt to flow and hence melt trapped in between the ridges. Hence, fiber-matrix mechanical interlocking with treated fibers is better.

3.2 Attenuated Total Reflectance Fourier Transform Infrared Spectroscopy

In order to investigate the possible changes in chemical composition of CF by plasma treatment, Attenuated total reflectance Fourier transform infrared spectroscopy (ATRF-TIR) analysis (Fig. 6) was done in mid infrared range (700–4,000 cm^{-1}). Spectrum of untreated fiber does not show any significant peaks

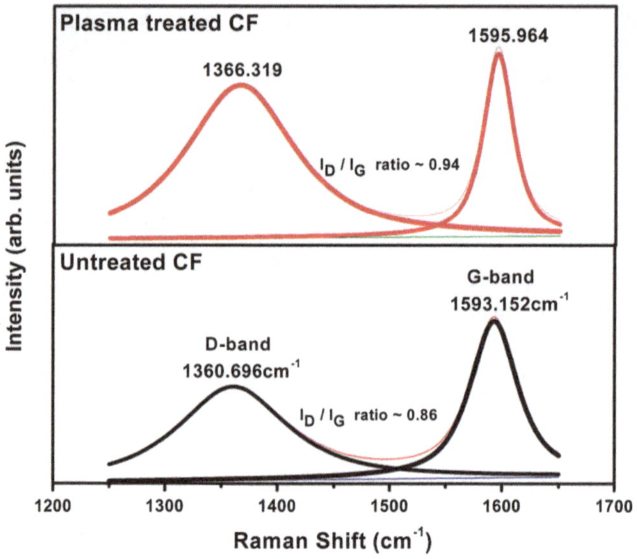

Fig. 7 Raman spectra of the untreated and plasma treated carbon fibers [22]

while that of plasma treated fibers, presence of oxygenated polar functional groups was observed. Ether, carboxyl and carbonyl groups were observed corresponding to wave number range 950–1,200, 1,650–1,710 cm^{-1}, respectively. These functional groups were responsible for improvement in adhesion between the matrix and fabric.

3.3 Micro Raman Spectroscopy

Carbon materials, such as carbon fibers and other sp^3 bonded amorphous carbons are strong Raman scatterers and the Micro Raman Spectroscopy (MRS) technique enables to distinguish between various structural organizations in these materials [25–27]. The first-order Raman spectra bands with peak positions 1,360 and 1,593 cm^{-1} are the main features of carbon materials and are called D bands (disordered) and G bands (graphitic), respectively [28–31].

The degree of structural disorder on the surface of CF due to CRNOP treatment was characterized by the ratio of integrated intensity of disorder induced (I_D) to Raman allowed band (I_G). The ratio I_D/I_G (Fig. 7) showed slight increase and a decrease in surface crystallite size (La). The size of crystallites located in the surface regions (La) was calculated using the empirical formula by Tuinstra and Koenig [32]; La for treated and untreated CF was 4.68 and 51.6 nm, respectively. During surface treatment graphitic microstructure of CF is partially destructed,

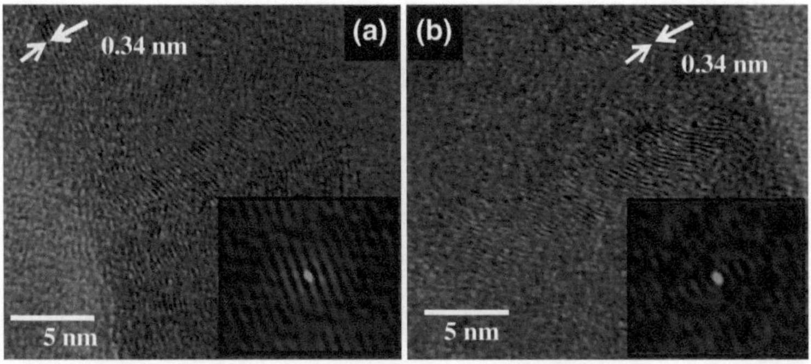

Fig. 8 HRTEM images of carbon fibers; **a** untreated, **b** plasma treated fibers; **c** and **d** are their corresponding auto correlated images, respectively [20]

the crystallite size is reduced, and the activity of the graphite crystallite boundary is improved [33, 34]. Figure 8 supports this by indicating the increased I_D/I_G ratio, hence induced distortion (reduced crystallinity) due to treatment to CF.

3.4 High Resolution Transmission Electron Microscopy

High Resolution Transmission Electron Microscopy (HRTEM) is an indispensable tool for examining the finer details of the fiber surface. It was preferred to compare the induced distortion in the graphitic planes of treated and untreated CF.

Figure 8 shows HRTEM images of longitudinal thin section for untreated and treated fibers. Both fibers have inter-planar spacing of 0.34 nm; typically observed for (210) planes of PAN based CF and high purity carbon and graphite materials [35–37]. The micrographs show the orientation of small graphite crystallites in CF. Both shows the coexistence of crystalline and amorphous phases which is accordance to the literature [38, 39]. Warner et al. [40] suggested that the structure of PAN fibers is constitutive of ordered and amorphous domains with the length of the ordered regions ranging from 80 to 100 Å, roughly twice that of the disordered regions. During fiber manufacturing process uneven distribution of stresses during the drawing step leads to the existence of both the phases [38]. The auto correlated images of small sections are shown in respective inserts. While comparing the inserts, the distorted graphitic plane can be easily seen in the case of treated CF which supports the results from Raman spectroscopic studies (I_D/I_G ratio found increased in case of treated CF). The planes are more smother and regular in case of untreated CF. Distortion can be correlated with pitting on CF and hence improves their compatibility with the matrix material (FESEM studies in Sect. 3.1).

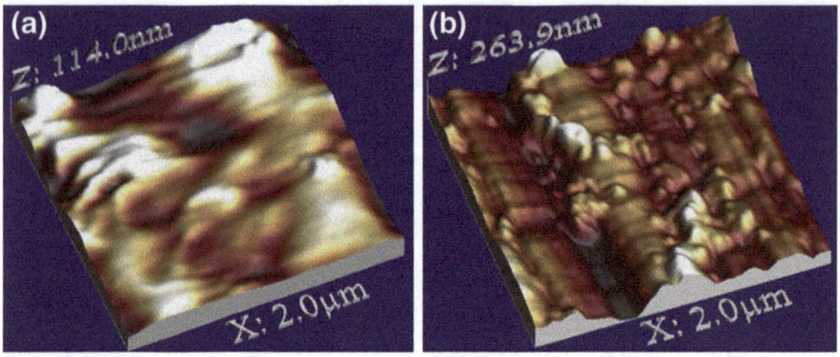

Fig. 9 AFM images of carbon fibers; **a** untreated **b** Plasma treated indicating increase in surface roughness from 114 to 264 nm [19, 20]

3.5 Atomic Force Microscopy

Atomic Force Microscopy (AFM) studies (Fig. 9) were carried out to analyze the topographical and morphological changes induced on the surface of CF due to plasma treatment. Fiber tows of untreated and treated CF were mounted separately on the stainless steel magnetic stubs. The fine striations on the untreated carbon fiber surface were due to the spinning of the fiber precursor [41]. The observed features similar to FESEM studies such as surface etching, increased perforation and presences of deeper and narrower ridges were more clearly seen on the treated fiber. The average surface roughness values for untreated and treated carbon fibers were 23.28 and 52.43 nm respectively, which confirmed the increase in surface area and alteration in morphology due to the treatment. The increase in surface roughness of treated carbon fibers is beneficial for enhance their reactivity towards matrix materials, since a rougher fiber topography would lead to a higher degree of mechanical interlocking between fiber and matrix [42].

3.6 Fiber: Matrix Adhesion Test

A simple test was performed to ensure enhancement in fiber-matrix adhesion due to treatment. A small sample of fibers (treated and untreated) was dipped n PES solution (in DCM 20 wt %) for 10 min followed by careful withdrawal of the fibers and drying in identical way. The difference in layer of matrix adhering to the fiber strand was examined with SEM (Fig. 10). Figure 10b confirms more polymer adhered to the treated fiber rather than the untreated one. As compare to untreated fibers (Sect. 3.1) deeper channels (along longitudinal axis) on the treated CF surface provide more surface areas (denier per filament) for polymer to adhere adequately, which in turns responsible for enhanced adhesion and hence the better mechanical interlocking between the fiber and matrix.

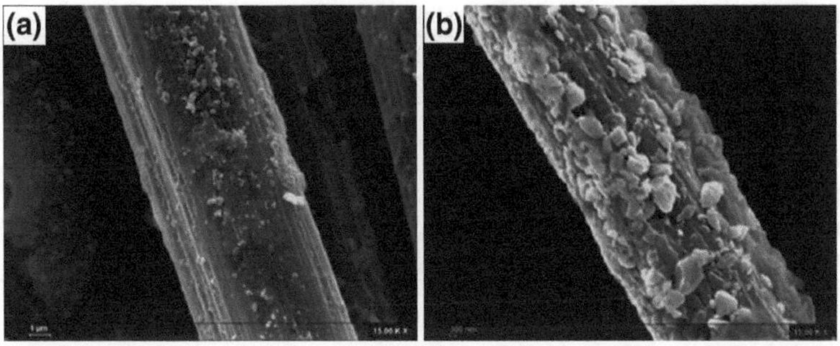

Fig. 10 SEM (×15 K) of impregnated fibers; **a** before treatment **b** after plasma treatment [20]

The analysis from all surface characterization techniques revealed that the plasma treatment on fiber surface altered its original inertness of and led to enhancement in fiber matrix adhesion, which resulted in improved performance properties of their composites as discussed in subsequent section.

3.7 Mechanical Strength of CF

The reduction in single fiber strength due to the fiber modification methods is a critical issue. Generally, fiber surface modification method increase the fiber-matrix interfacial strength but at the cost of decremented single fiber strength properties. The CRNOP treatment reduced the strength of a carbon fiber (5 to 10 %) [22].

4 Characterization of Developed Composites

The composites developed were characterized for their physical, mechanical and tribological properties as discussed in following sections.

4.1 Characterization of Composites with Treated and Untreated CF

Table 1 shows the details and designations of composites with tailored surfaces while Table 2 summarizes the properties of composites and positive changes due to CRNOP treated fabric in a composite. The reinforcement influenced the heat distortion temperature of PES appreciably (25–30 °C). However, the CF treatment

Table 1 Details of unmodified and PTFE modified PES-CF composites [19, 20]

Designations of composites[a]	Av. PTFE particle size (FESEM studies) (nm)	Shape of PTFE fillers
PES_{CFU}	–	–
PES_{CFT}	–	–
PES_{CFTN}	50–80	Highly spherical
PES_{CFTSM}	200–250	Sub rounded
PES_{CFTM}	400–450	Sub angular

[a]PES_{CFU}—Composites with untreated CF
PES_{CFT}—Composites with treated CF
PES_{CFTN}—Composite with treated CF and nano- sized (50–80 nm) PTFE on the surface
PES_{CFTSM}—Composite with treated CF and submicron sized (150–200 nm) PTFE on the surface
PES_{CFTM}—Composite with treated CF and micron sized (400–450 nm) PTFE on the surface

Table 2 Physical and mechanical properties of CF-PES composites reinforced with untreated and plasma treated CF [20]

Properties/materials	PES	PES_{CFU}	PES_{CFT}	% changes due to CF treatment
Fiber weight (%)	–	67.50	68.24	–
Void fraction	–	0.47	0.37	↓ 21.3
Density (g/cm^3)	1.37	1.52	1.54	↑ 1.3
HDT ASTM D648 (°C)	204	227	233	↑ 2.6
Tensile strength (MPa)	84	744	778	↑ 4.4
Tensile modulus (GPa)	3	65	76	↑14.4
Toughness (MPa)	60	4.1	4.3	↑4.7
Flexural strength (MPa)	112	692	835	↑17.1
Flexural modulus (GPa)	2.8	54	68	↑20.6
Interlaminar shear strength (MPa)	–	36	46	↑21.7

had almost negligible effect on the HDT values of the composites (1.5–2.5 %). Composites containing plasma treated CF proved superior to those with untreated CF confirming the improved fiber-matrix adhesion as a result of treatment due to the increment in fiber matrix mechanical interlocking.

4.2 Tribo-Characterization of Composites

Performance of composites was evaluated in adhesive sliding wear mode using pin on disc configuration as discussed in following subsection.

4.2.1 Methodology for Tribo-Evaluation of Composites

Tribological studies in adhesive wear mode were carried out on UMT-3MT Tribometer supplied by CETR, USA. Prior to the experiment the composite pin slid against a rough mild steel disc for uniform contact. Initial weight of the pin was measured after cleaning ultrasonically with petroleum ether followed by drying. The pin was slid against a mild steel disc (Ra values range 0.1–0.2 μm) at a constant speed of 1 m/s. After the experiment, pin was again weighed with an accuracy of 0.0001 g and weight loss readings were used to calculate the specific wear rate (K_o) of composites. μ as a function of time during sliding was recorded with the help of viewer software.

The specific wear rate (K_o) was calculated using the equation:

$$K_0 = \frac{W}{\rho L d}\ m^3 N^{-1} m^{-1} \qquad (1)$$

Where; W is the weight loss in kg, ρ the density of pin in kg/m^3, L the load in N and d the sliding distance in meters. The experiment was repeated for three times and the average of two closest values of weight loss was used for specific wear rate calculations.

4.2.2 Tribological Aspects of PTFE as a Solid Lubricant

PTFE has a peculiar morphological and molecular structure and has a high molecular weight inert fluorocarbon compound which demonstrates mitigated London dispersive forces due to highly electronegative F- atoms. In PTFE molecule, C–F forms non-reactive and instantaneous polarized multi poles. With the increases in surface contact, the polarizability increases due to the dispersed electron clouds hence closer interaction between different molecules. Tribological point of view, this is the most exploited solid lubricant in various amounts and sizes in the bulk of the composites barring nano-size, in general. In PTFE fluorine atoms are close enough to form a smooth cylindrical surface against which other molecules can easily slide. At larger scale, the long chains of PTFE orient on the counter face surface during sliding creating a fine coherent transfer film. The transfer film creates a low shear-strength interface with the bulk PTFE material [43]. Hence the interaction is between PTFE film and the PTFE in composite leading to least possible adhesion and hence very low μ. This film transferring ability depends on the size and amount of PTFE particles apart from operating conditions.

4.2.3 Tribo-Characterization of Composites

The essence of performance of composites is shown in Table 3 which elaborates on the influence of two modifications (plasma treatment to the CF and PTFE on the surface of a composite) on W_R wear resistance and μ. Overall W_R of the composites was in the range 4.8 to 7.8 × 10^{14} Nm/m^3, which is rated as quite high and μ was in the range 0.06–0.25 which is a desirable range for such composites. In all the cases with increase in load W_R and μ decreased appreciably and these are the general trends reported in the literature.

Table 3 The essence of CF-PES composites performance due to treatments (plasma treatment to the CF and PTFE on the surface of a composite) [20]

Composites	Limit load* (N)	Wear resistance $(W_R)x10^{14}$ Nm/m^3 under load (×100 N)						Coefficient of friction (μ) under load (×100 N)					
		2	4	6	7	8	9.5	2	4	6	7	8	9.5
PES_{CFU}	700	6.5	6.4	5.6	5.2	F	F	0.25	0.19	0.18	0.14	F	F
PES_{CFT}	700	7.2	7.1	6.0	5.3	F	F	0.21	0.18	0.13	0.12	F	F
PES_{CFTM}	800	7.5	7.2	6.4	5.7	5.0	F	0.16	0.13	0.11	0.11	0.09	F
PES_{CFTSM}	900	7.6	7.3	6.5	5.4	5.1	F	0.15	0.12	0.10	0.10	0.08	F
PES_{CFTN}	950	7.8	7.4	6.6	5.6	5.4	4.8	0.14	0.11	0.09	0.09	0.07	0.06

*Shows failure of composite under the selected load

Influence of Plasma Treatment on Fibers

The plasma treatment led to increase in W_R and decreased μ (Table 3) of CF-PES composites which is a significant improvement. As compared to the untreated ones, μ values of treated CF-PES composite lowered by ~15 % with almost equal increment in wear resistance. The treatment to the fibers led to the increase in the surface roughness and inclusion of functional groups which resulted in more fiber-matrix adhesion as discussed in the earlier section; leading to more resistance to peeling off or breakage of fibers during sliding and hence lower wear.

Influence of PTFE Modification on the Surface and Size of Particles

The μ and W_R were highly influenced due to PTFE modification rather than size of PTFE particles on the surface. PTFE inclusion improved the μ and W_R of composites to ~33 and ~14 % respectively. Table 3 summarizes the trends in improvement (Fig. 11) due to various PTFE modifications on Tribo-performance parameters (μ, W_R and limiting load).

The incremented limiting load value has established the efficacy of PTFE surface modification, the limiting load for PES_{CFU} and PES_{CFT} composites was up to 700 N, while for tailored composites it was from 800 to 950 N. For PES_{CFTN} composite (tailored with 50–80 nm size PTFE); limiting load value was 950 N with $W_R \approx 4.8 \times 10^{14}$ Nm/m^3 and μ value ≈ 0.06 confirming potential of nano-PTFE.

Sliding wear performance of the composites improved with decrement in the size of PTFE fillers. Well spherical nano fillers provide high interfacial area between the fillers and matrix. This leads to a better bonding between the two phases and hence better strength and toughness properties [6, 44]. Topographical smoothening and a rolling effect due to the inclusion of nano-fillers at the surface is the reason for improved friction and wear performance of PES_{CFTN} composite. It is of utmost importance that the NPs should be uniformly dispersed to get the best property profile. To avoid agglomeration the minimum wt % of fillers is to be

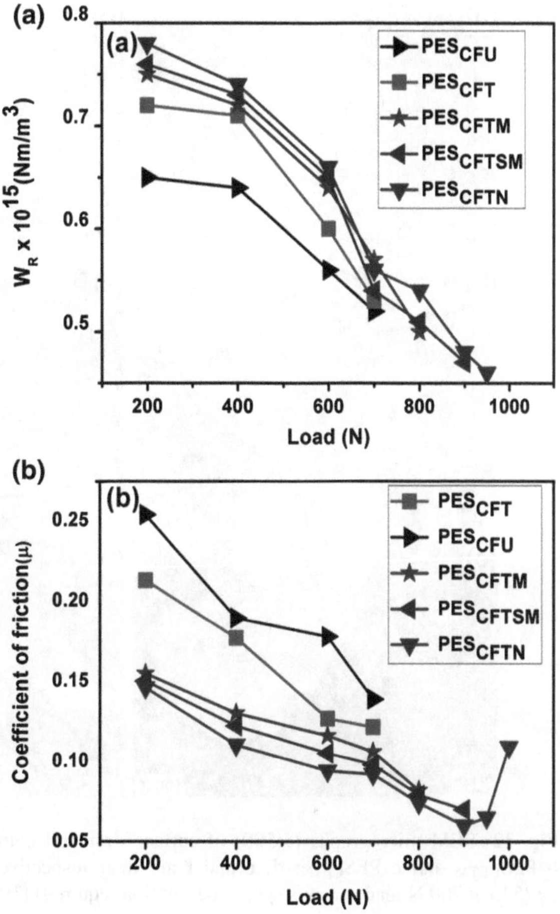

Fig. 11 **a** Coefficient of friction; **b** specific wear rates as a function of increasing load for all surface designed series of composites [19, 20]

used. In literature on polymeric NCs generally 2–3 wt % of nano-fillers is claimed [45–48] to be the optimum amount. The main feature of PTFE NPs which influence the wear performance is their huge interfacial surface area.

It was desired to see the service life of the designed surfaces. The CF-PES composites (without and with PTFE particles at the surface) were slid against steel disc till the steady state friction value starts fluctuating at high friction torque. This signposted the limiting life of modified composites when few or no PTFE particles left on its surface for replenishment of transferred film on the steel disc.

Higher the limiting time, more is the tribo-utility of the surface. In the long experiment at 700 N load, the limiting sliding times for the composite surface without PTFE, with micro sized PTFE and nano—PTFE were; 8.33, 17.7 and 21.3 h respectively; indicating beneficial effects of PTFE on the composite surface [19]. For PES_{CFTN} composites, the transfer of thin tenacious PTFE film on the steel disc surface was effective in maintaining steady μ values for long time.

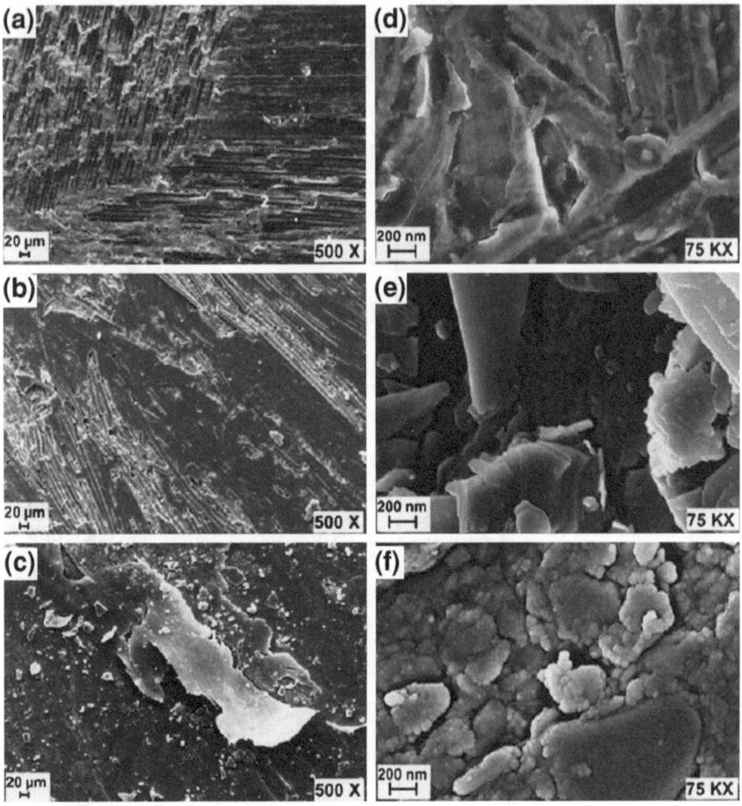

Fig. 12 SEM micrographs (×500) of surface designed composites after wearing; **a** PES_{CFT}, **b** PES_{CFTM} and **c** PES_{CFTN}; **d**, **e** and **f** are their respective high resolution FESEM images (×75 k) at 700 N load at 1 m/s speed (permission required) [19, 20]

Figure 12 shows SEM and FESEM micrographs of worn surfaces of surface designed composites. The fibers were fully covered with the nano PTFE fillers. PTFE layer is efficiently transferred on the counter surface and sliding is between PTFE layer on the composite surface and thin tenacious layer transferred on the counter surface. The existence on nano-fillers (Fig. 12f) and micro scale (Fig. 12e) is clearly visible with high resolution FESEM images of worn composites surface, which were absent for composites without tailored surface (Fig. 12d).

5 Concluding Remarks

Surface designing of PES_{CFT} composites with micro, sub-micro and nano-scale PTFE fillers improved the overall tribo-performance of composites; though the reduction in μ was significant rather than the wear resistance. The idea of surface treatment of a composite with solid lubricants to safeguard the composites from an un-intentional reduction in strength properties and increase in the cost (if solid lubricant is employed in the whole composite rather than the surface both factors viz. strength and economics are affected significantly) proved successful. Surface designing enhanced limiting load values of composites significantly from 700 to 950 N, limiting running time from 8 to 21 h; reduction in μ (from 0.12 at 700 N load to 0.065 at 900 N load) and W_R (from 5.2 to 5.6×10^{14} Nm/m^3 at 700 N load) especially in the case of PES_{CFTN} composite. The increased surface area of contact due to the inclusion of nano—PTFE at the composite surface was responsible for enhanced tribo-performance of PES_{CFTN} composite.

References

1. Donnet C, Erdemir (2001) A Historical developments and new trends in tribological and solid lubricant coatings. Surf Coat Technol 180:76–84
2. Teer DG (2001) New solid lubricant coatings. Wear 251:1068–1074
3. Blanchet TA, Kennedy FE (1992) Sliding wear mechanism of Polytetrafluoroethylene (PTFE) and PTFE composites. Wear 153:229–243
4. Nejhad MNG, Veedu VP, Yuen A, Askari D (2007) Polymer matrix composites with nanoscale reinforcements. Patent US 2007(0142548):A1
5. Bijwe J, Hufenbach W, Kunze K, Langkamp A (2008) Polymeric composites and bearings with engineered tribo-surfaces. In: Schlarb AK, Friedrich K (eds) Tribology of polymeric nano-composites. Elsevier, Amsterdam
6. Friedrich K, Zhang Z, Schlarb AK (2005) Effects of various fillers on the sliding wear of polymer composites. Compos Sci Technol 65:2329–2343
7. Friedrich K, Fakirov S, Zhang Z (2005) Polymer composites—from nano to macro scale, Springer, US
8. Chang L, Zhang Z, Ye L, Friedrich K (2008) Synergistic effect of nanoparticles and traditional tribo-fillers on sliding wear of polymeric hybrid composites. In: Schlarb AK, Friedrich K (eds) Tribology of polymeric nano-composites. Elsevier, Amsterdam
9. Zhang MQ, Rong MZ, Friedrich K (2006) Wear resisting polymer nanocomposites: reparation and properties. In: Mai YW, Yu ZZ (eds) Polymer nanocomposites. CRC Press, Washington
10. Zhang ZZ, Su FH, Wang K, Jiang W, Men XH, Liu WM (2005) Study on the friction and wear properties of carbon fabric composites reinforced with micro- and nano-particles. Mater Sci Eng A 404:251–258
11. Su FH, Zhang ZZ, Guo F, Wang K, Liu WM (2006) Effects of solid lubricants on friction and wear properties of nomex fabric composites. Mater Sci Eng, A 424:333–339
12. Kuo MC, Tsai CM, Huang JC, Chen M (2005) PEEK composites reinforced by nano-sized SiO_2 and $Al2O_3$ particulates. Mater Chem Phys 90:185–195
13. Rong MZ, Zang M, Liu H, Zeng HM, Wetzel B, Freidrich K (2001) Microstructure and tribological behavior of polymeric nanocomposites. Ind Lubr Tribol 53:72–77
14. Burris DL, Boes B, Bourne GR, Sawyer WG (2007) Polymeric nanocomposites for tribological applications. Macromol Mater Eng 292:387–402
15. Friedrich K, Lu Z, Hager AM (1993) Overview on polymer composites for friction and wear application. Theoret Appl Fract Mech 19:1–11
16. Dwyer-Joyce RS, Sayles RS, Ioannides E (1994) An investigation into the mechanisms of closed three-body abrasive. Wear 175:133–142

17. Bahadur S (2000) The development of transfer layers and their role in polymer tribology. Wear 245:92–99
18. Vaziri M, Spurr RT, Stott FH (1988) An investigation of the wear of polymeric materials. Wear 122:329–342
19. Sharma M, Bijwe J (2012) Surface engineering of polymer composites with nano and micron sized PTFE fillers. J Mater Sci 47:4928–4935
20. Sharma M (2011) Carbon fabric reinforced polymer composites: development, surface designing by micro and nano PTFE and performance evaluation, Ph D thesis, Indian Institute of Technology, Delhi, India
21. Tiwari S, Bijwe J, Panier S (2011) Tribological studies on Polyetherimide composites based on carbon fabric with optimized oxidation treatment. Wear 271:2252–2260
22. Tiwari S, Sharma M, Panier S, Mutel B, Mitschang P, Bijwe J (2011) Influence of cold remote nitrogen oxygen plasma treatment on carbon fabric and its composites with specialty polymers. J Mater Sci 46:964–974
23. Tiwari S, Bijwe J, Panier S (2011) Role of Nano-YbF3 treated carbon fabric on improving abrasive wear performance of polyetherimide composites. Tribol Lett 42:293–300
24. Tiwari S, Bijwe J, Panier S (2011) Polyetherimide composites with gamma irradiated carbon fabric: studies on abrasive wear. Wear 270:688–694
25. Meyer N, Marx G, Brzezinka KW (1994) Raman spectroscopy of carbon fibres. J Anal Chem 349:167–168
26. Sarraf H, Škarpová L, Louda P (2007) Surface modification of carbon fibers. J Achievements Mater Manuf Eng 25:24–30
27. Montes-Morán MA, Young RJ (2002) Raman spectroscopy study of high-modulus carbon fibres: effect of plasma-treatment on the interfacial properties of single-fiber-epoxy composites part II: characterisation of the fibre-matrix interface. Carbon 40:857–875
28. Morita K, Murata Y, Ishitani A, Murayama K, Ono T, Nakajima A (1986) Characterization of commercially available PAN (polyacrylonitrile)-based carbon fibers. Pure Appl Chem 58:455–468
29. Sharma SP, Lakkad SC (2011) Effect of CNTs growth on carbon fibers on the tensile strength of CNTs grown carbon fiber-reinforced polymer matrix composites. Compos Part A 42:8–15
30. Liu Y, Pan C, Wang J (2004) Raman spectra of carbon nanotubes and nanofibers prepared by ethanol flames. J Mater Sci 39:1091–1094
31. Nemanich RJ, Solin SA (1979) First and second order Raman scattering from finite-size crystals of graphite. Phys Rev B 20:392–401
32. Tuinstra F, Koenig JL (1970) Raman spectrum of graphite. J Chem Phys 53:1126–1130
33. Chaudhuri SN, Chaudhuri RA, Benner RE, Penugonda MS (2006) Raman spectroscopy for characterization of interfacial debonds between carbon fibers and polymer matrices. Compos Struct 76:375–387
34. Cao WW, Zhu B, Jing M, Wang CG (2008) Raman spectra of PAN-based carbon fibers during surface treatment. NCBI PubMed 28:2885–2889
35. Tressaud A, Chambon M, Gupta V, Flandrois S, Bahl OP (1995) Fluorine-intercalated carbon fibers III. A transmission electron microscopy study. Carbon 33:1339–1345
36. Sharma SP, Lakkad SC (2009) Morphology study of carbon nanospecies grown on carbon fibers by thermal CVD technique. Surf Coat Technol 203:1329–1335
37. Gaskell DR (1981) Introduction to the thermodynamics of materials. McGraw-Hill, New York
38. Bai YJ, Wang CG, Lun N, Wang YX, Yu MJ, Zhu B (2006) HRTEM microstructures of PAN precursor fibers. Carbon 44:1773–1778
39. Hai XS, Ying ZF, Huan LS, Dong-mei H, Qing-yun C (2010) Catalytic graphitization of Mo-B-doped polyacrylonitrile (PAN)-based carbon fibers. J Central South Universityn Technol 17:703–707
40. Warner SB, Uhlmann DR, Peebles LH (1975) Ion etching of amorphous and semicrystalline fibers. J Mater Sci 10:758–764

41. Zhang X, Huang Y, Wang T (2006) Surface analysis of plasma grafted carbon fiber. Appl Surf Sci 253:2885–2892
42. Paredes JI, Alonso AM, Tascón JMD (2000) Atomic force microscopy investigation of the surface modification of highly oriented pyrolytic graphite by oxygen plasma. J Mater Chem 10:1585–1590
43. Schadler LS, Brinson LC, Sawyer WG (2007) Polymer nanocomposites: a small part of the story. J Miner Met Mater Soc 59:53–60
44. Kocsis JK, Zhang Z (2005) Structure-property relationships in nanoparticle/semi-crystalline thermoplastic composites. In: Calleja JFB, Michler G (eds) Mechanical properties of polymers based on nanostructure and morphology. CRC Press, New York
45. Xian G, Zhang Z, Friedrich K (2006) Tribological properties of micro and nanoparticles-filled Poly(etherimide) Composites. J Appl Polym Sci 101:1678–1686
46. Chang L, Zhang Z, Zhang H, Friedrich K (2005) Effect of nanoparticles on the tribological behaviour of short carbon fibre reinforced Poly(etherimide) composites. Tribol Int 38:966–973
47. Zhang ZZ, Sua FH, Wang K, Jiang W, Mena XH, Liu WM (2005) Study on the friction and wear properties of carbon fabric composites reinforced with micro- and nano-particles. Mater Sci Eng, A 404:251–258
48. Park DC, Kim SS, Kim BC, Lee SM, Lee DG (2006) Wear characteristics of carbon-phenolic woven composites mixed with nano-particles. Compos Struct 74:89–98

Tribology of MoS_2-Based Nanocomposites

Kunhong Hu, Xianguo Hu, Yufu Xu, Xiaojun Sun and Yang Jiang

Abstract In this chapter, the preparation and tribological properties of MoS_2-based nanocomposites were reviewed, including nanocomposites of MoS_2 with different morphologies, MoS_2/inorganic compound nanocomposites, MoS_2/polymer nanocomposites, and Ni–P–(nano-MoS_2) composite coatings. The nanocomposites of MoS_2 can be prepared by mechanical-mixing two kinds of nano-MoS_2 with different morphologies or chemically synthesizing from sodium molybdate and different sulfides. The nanocomposites of MoS_2 reveal better tribological properties than their original materials. Moreover, the chemical method presents advantages over the mechanical one in the preparation of the MoS_2 nanocomposites with different morphologies for lubrication applications. Using an appropriate chemical method may produce MoS_2/inorganic compound nanocomposites such as MoS_2/TiO_2 nanocomposite. Two kinds of nanoparticles (nano-MoS_2 and nano-TiO_2) reveal a synergistic effect on the tribological properties of the MoS_2/TiO_2 nanocomposite. MoS_2/polymer nanocomposites may be prepared by adding nanosized MoS_2 into polymers or using the chemical intercalation technology. The chemical intercalation technology may lead to disperse MoS_2 into polymer matrix better than the mechanical-filled way. However, the intercalation compound of MoS_2/polymer can not present a satisfactory lubrication performance, because the intercalation process destroys the 2H structure of MoS_2 with better lubricity. The Ni–P coatings may be co-deposited with nanosized MoS_2 on medium carbon steel substrate by electroless plating. The

K. Hu (✉)
Department of Chemical and Materials Engineering, Hefei University,
Hefei 230601, People's Republic of China
e-mail: hukunhong@163.com

X. Hu (✉) · Y. Xu
Institute of Tribology, Hefei University of Technology, Hefei 230009,
People's Republic of China
e-mail: xghu@hfut.edu.cn

X. Sun
Stat Key Laboratory, Lanzhou Institute of Chemical Physics, Chinese Academy of Sciences,
Lanzhou 730000, People's Republic of China

Y. Jiang
School of Materials Science and Enginering, Hefei University of Technology,
Hefei 230009, People's Republic of China

J. Paulo Davim (ed.), *Tribology of Nanocomposites*, Materials Forming,
Machining and Tribology, DOI: 10.1007/978-3-642-33882-3_3,

obtained Ni–P–nano-MoS_2 composite coating shows an excellent lubricating performance. The present review concluded the synthesis and tribological applications of MoS_2-based nanocomposite well.

1 Introduction

Nanocomposites have wide applications in modern materials science and nanotechnology. Recently, the significance of nanocomposites in tribology was also paid so much attention. The nanocomposites may be prepared by mechanical mixing, chemical synthesis and coating technology. The components in a nanocomposite may offset their defects and enhance their merits mutually. Thus, the nanocomposites usually have better performances in friction reduction and wear resistance than their original materials. Some solid lubricants, such as molybdenum disulfide (MoS_2), graphite, and carbon nanotube, are often used as materials to synthesize nanocomposites. Herein, several selected features concerning the MoS_2-based nanocomposites were reviewed based on our recent researches and results reported by other researchers. In the second section, the structure and properties of bulk 2H-MoS_2 were reviewed. Section 3 describes the development in nanosized MoS_2 (nano-MoS_2). Section 4 is focused on the synthesis and tribological properties of MoS_2-based nanocomposites, including MoS_2 nanocomposites with different morphologies, MoS_2/inorganic compound nanocomposites, MoS_2/polymer nanocomposites, and Ni–P–(nano-MoS_2) composite coatings.

2 Molybdenum Disulfide

Molybdenum disulfide (MoS_2) is the main component of molybdenite that is the principal ore of molybdenum. MoS_2 has three crystal states, i.e. 1T, 2H, and 3R [1]. The 2H layered crystal structure is usually considered as the most important factor for lubrication of MoS_2. The commercial lubricant of bulk 2H-MoS_2 presents a platelet-like shape (Fig. 1) [2]. The bulk 2H-MoS_2 is composed of layered structures that contains strong S–Mo–S covalent bonds in inside layers and weak Van der Waals gaps between molecular layers. The 2H layered structure results in a strong (002) peak in the powder X-ray diffraction pattern (XRD) of MoS_2 (Fig. 2) [3].

The Van der Waals gaps between MoS_2 layers are easy to slide under the friction shearing (Fig. 3). In addition, S atoms on MoS_2 have an intensive adsorption effect on the metal surface. The two characteristics may provide persistent lubrication for metal friction pairs, especially in extreme environments such as high-temperature and high-vacuum [4, 5]. Thus, MoS_2 has become an important solid lubricant in aviation and aerospace. Moreover, MoS_2 is also a known-well additive in lubricating oils, polymers, and coatings [1, 6, 7].

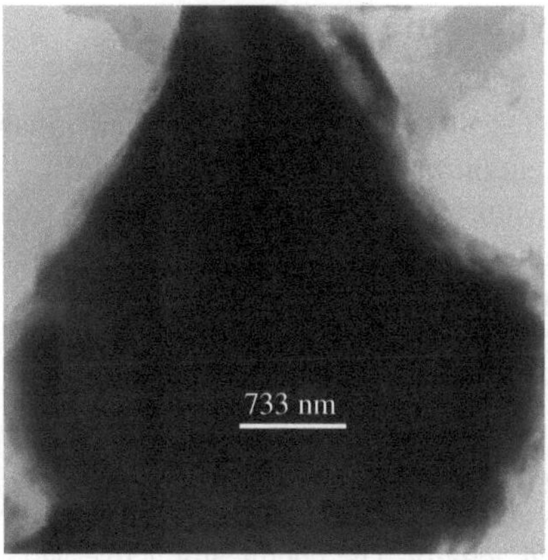

Fig. 1 TEM image of 2H-MoS2 (adapted from Ref. [2])

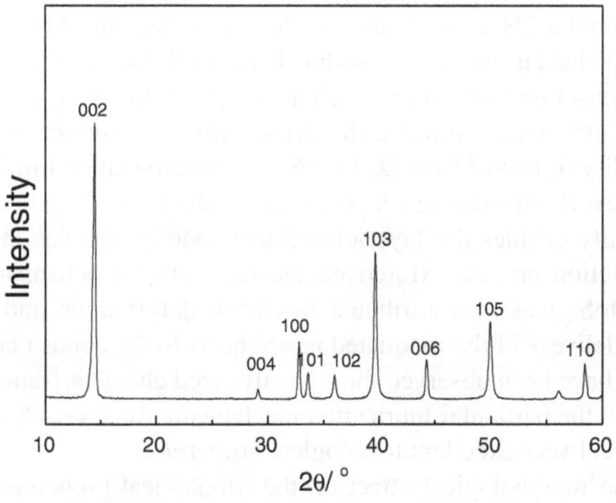

Fig. 2 X-ray diffraction pattern of 2H-MoS_2 powder (adapted from Ref. [3])

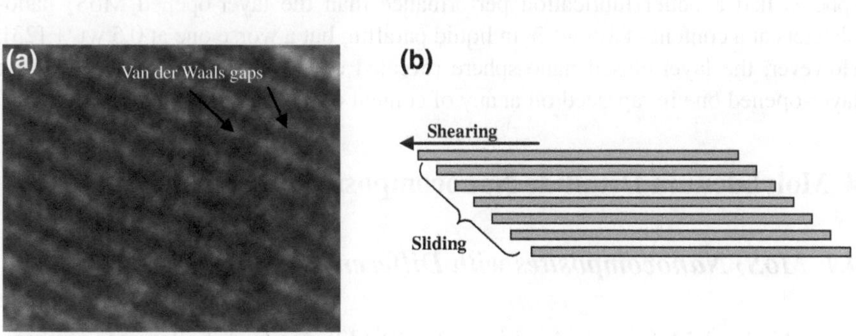

Fig. 3 Van der Waals gaps between MoS_2 layers (**a**) and schematic diagram of their shearing-sliding (**b**)

3 Nanosized Molybdenum Disulfide

Nano-MoS_2 usually presents better lubrication performance than bulk MoS_2. Thus, considerable attention has been given to nano-MoS_2. There has been a lot of researches on the synthesis [8–13] and tribology of nano-MoS_2 [14–26]. The chemical routes to synthesize nano-MoS_2 include gas phase growth [8], hydrothermal or solvothermal synthesis [10], decomposition of precursors [11, 12], etc. The synthesized nano-MoS_2 involves tube-like [8, 9], platelet-like [11], sphere-like [12] and fullerene-like [8] shapes. The morphologies of nano-MoS_2 can be categorized into two: layer-opened and layer-closed.

The layer-opened MoS_2, such as platelet-like nano-MoS_2 (MoS_2 nano-platelet), contains basal surfaces and rim-edge surfaces [27]. The atoms on the rim-edge surface have high chemical activity. The chemically active MoS_2 nano-platelet is easy oxidized during friction process. The oxidation resultants, such as MoO_3 and sulfates, may function as a lubrication film to reduce friction [28, 29]. However, an excessive oxidation can also weaken the lubrication of nano-platelet. Because MoS_2 nano-platelet has a similar 2H layered structure to that of the bulk MoS_2, its lubrication may also be explained using the easy sliding between S–Mo–S molecular layers [23].

Forming layer-closed structures, such as inorganic fullerene-like, tube-like and hollow sphere-like, may eliminate the active rim-edge surface and increase the chemical stability of nano-MoS_2 [2, 15, 16, 23]. The oxidation film is not the main reason for the excellent tribological properties of the layer-closed nano-MoS_2. The chemical stability enables the layer-closed nano-MoS_2 to function as lubrication well during friction process. Moreover, the lubrication mechanism of the layer-closed nano-MoS_2 was also attributed to elastic deformation and exfoliation of MoS_2 and the delivery of the exfoliated nano-sheets to the contact area [14, 15, 24, 30, 31], which have been observed through advanced characterization technologies [25, 26]. Due to the particular lubrication mechanism, the layer-closed nano-MoS_2 can usually reveal very excellent tribological properties.

Recently, the morphological effect on the tribological properties of MoS_2 was studied in liquid paraffin (LP) and rapeseed oil. The layer-closed MoS_2 nano-spheres had a better lubrication performance than the layer-opened MoS_2 nano-platelets at a content of 1.5 wt % in liquid paraffin, but a worse one at 0.5 wt % [23]. However, the layer-closed nano-sphere revealed considerable advantages over the layer-opened one in rapeseed oil at any of content used (unpublished results).

4 Molybdenum Disulfide Nanocomposites

4.1 MoS_2 Nanocomposites with Different Morphologies

Three kinds of MoS_2, namely, micro-platelet (325 meshes), hollow nano-sphere, and nano-platelet, were used to prepare MoS_2 nanocomposites with different morphologies by mechanical mixing [32]. The diameters of MoS_2 hollow

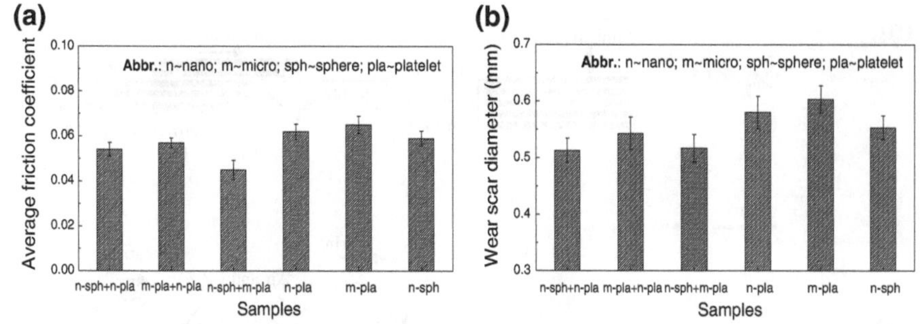

Fig. 4 Tribological properties of MoS_2 nanocomposites prepared by mechanical mixing (adapted from Ref. [32])

nano-spheres vary from 80 to 200 nm with an about 15 nm shell. The thickness of nano-platelet is about 7 nm and the length about 40 nm. Some composites were obtained by proportionally mixing any two of the three kinds of MoS_2 in liquid paraffin.

Figure 4 provides results of four-ball tribological tests for the 1.5 wt % MoS_2 nanocomposites in liquid paraffin [32]. The tests were conducted at 1450 rpm and 300 N for 30 min. Figure 4a shows the average friction coefficients of different nanocomposites. The pure MoS_2 nano-spheres presented better anti-friction performance in liquid paraffin than the two pure platelets-like MoS_2. However, the MoS_2 nanocomposites had lower friction coefficients than that of the pure MoS_2 nano-spheres in liquid paraffin. The lowest friction coefficient occurred in the LP sample with the nano-sphere/micro-platelet composite (20 wt % nano-spheres and 80 wt % micro-platelets). Thus, forming nanocomposites may improve the anti-friction performance of MoS_2.

Figure 4b provides the anti-wear results (average wear scar diameter) of four-ball tests. As shown in the figure, The LP sample with MoS_2 nano-spheres presented better anti-wear properties than that with MoS_2 micro-platelets or nano-platelets. Some of LP samples with the nanocomposites presented better anti-wear performances than that with any of the three pure MoS_2. The nano-sphere/nano-platelet nanocomposite (60 wt % nano-spheres and 40 wt % micro-platelets) presented the best anti-wear performance.

These mentioned above indicate that the morphology of MoS_2 has an influence on the tribological properties of MoS_2 nanocomposites. The nanocomposites of MoS_2 with different morphologies may improve the wear resistance and friction reduction of LP more than any of the three morphologies of MoS_2 singly did. The different tribological properties of the three kinds of MoS_2 were attributed to their different lubrication mechanisms. The lubrication mechanism of bulk MoS_2 is associated with the sliding between molecular layers induced by the friction shearing. With a similar layered structure to that of bulk 2H-MoS_2, MoS_2 nano-platelets may also present the shearing and sliding lubrication mechanism (Fig. 5a) [23].

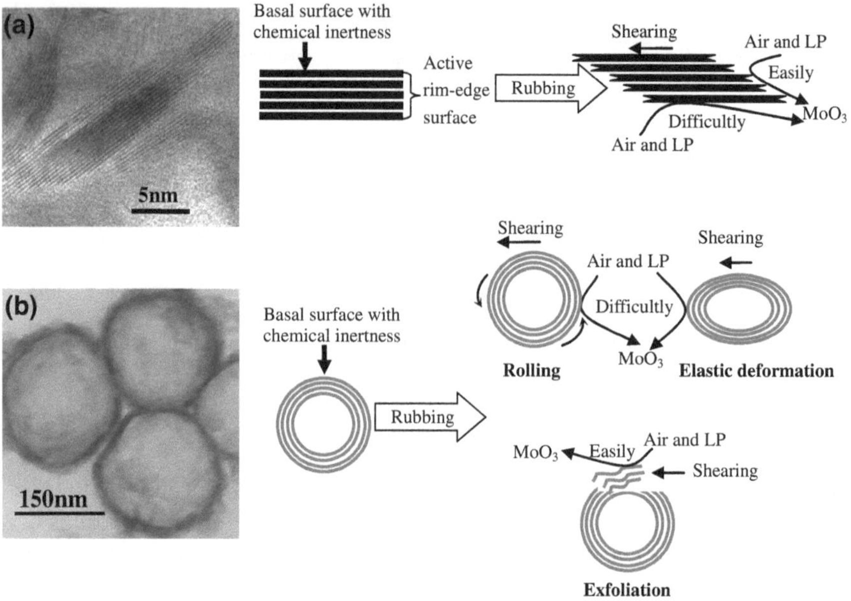

Fig. 5 Schematic of lubrication-wear mechanism of **a** MoS_2 nano-platelet and **b** MoS_2 nano-sphere (adapted from Ref. [23])

The excellent tribological properties of spherical nano-MoS_2 may be explained by its chemical inertness, rolling friction, deformation, and exfoliation-delivery of MoS_2 sheets to the contact area (Fig. 5b).

According the results of Stribeck curves [23], the rotation speed used (1450 rpm) fell in the end of the mixed lubrication. Thus, the oil film thickness between the friction pairs should be slightly larger than the surface roughness of friction pairs (0.032 μm). The MoS_2 nano-platelets with the smallest sizes easily penetrated into the friction contact region to function as lubrication. However, it was easy for the active nano-platelets to be excessively oxidized into MoO_3 (Fig. 5a). Thus, the nano-platelets didn't present better lubrication properties than the nano-spheres.

The better tribological properties of the nanocomposites resulted from the cooperation between two different lubrication mechanisms [32]. The size of the bulk MoS_2 micro-platelets exceeded the thickness of the oil film between the friction pairs. The adsorbed micro-platelets mainly functioned as a separation body between the friction pairs. Thus, the thickness of the oil film was magnified. The nano-MoS_2, i.e. nano-sphere or nano-platelet, was easier to penetrate into the contact region (Fig. 6a). When the micro-platelets adsorbed were worn by the friction shearing, the size of micro-platelet was close to the thickness of oil film. Then it occurred that the cooperation between the shearing-sliding of 2H structure and the rolling-deformation-exfoliation of nano-spheres (Fig. 6b). The similar cooperation

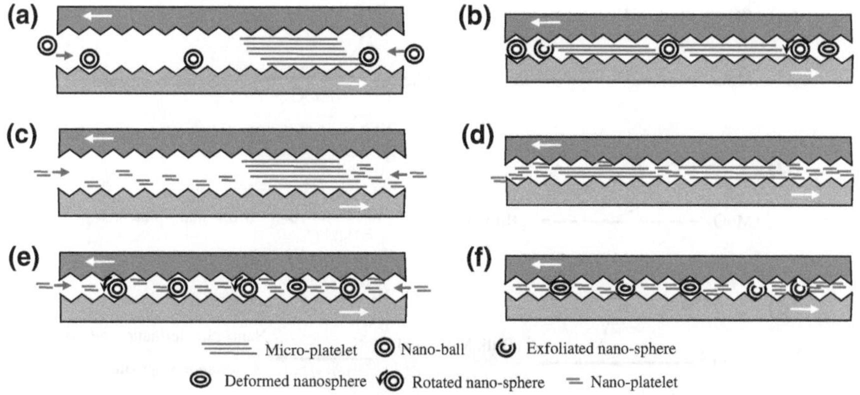

Fig. 6 Schematic of the synergistic lubrication between two kinds of MoS_2 particles: **a** initial stage and **b** stable stage lubricated by nano-spheres and micro-platelet, **c** initial stage and **d** stable stage lubricated by nano-platelets and micro-platelet, **e** initial stage and **f** stable stage slices lubricated by nano-spheres and nano-platelets (adapted from Ref. [32])

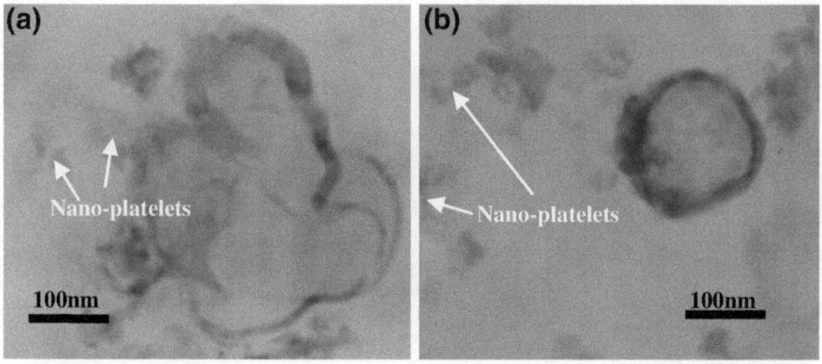

Fig. 7 TEM images of MoS_2 nanocomposite synthesized at a molar ratio (Na_2S to CH_3CSNH_2) of: **a** 1:2 and **b** 1:4 (adapted from Ref. [34])

was observed between nano-platelet and micro platelet or nano-sphere and nano-platelet (Figs. 6c–f).

MoS_3 may be synthesized using the reaction of sulfides and sodium molydate. Nano-MoS_2 can be obtained after heating MoS_3 in H_2 or N_2. The morphology of nano-MoS_2 is affected by the sulfides used [11, 33]. CH_3CSNH_2 (TAA) may produce spherical nano-MoS_2 while Na_2S platelet-like one. It was possible to prepare MoS_2 nanocomposite with different morphologies by adjusting the proportion of the two sulfides [34]. However, Na_2S can disturb the forming processes of nano-spheres especially at low dosages of TAA (Fig. 7a). The nano-sphere/nano-platelet composite can be obtained only at high TAA dosages (Fig. 7a). The Schematic of forming MoS_2 nanocomposite was shown in Fig. 8.

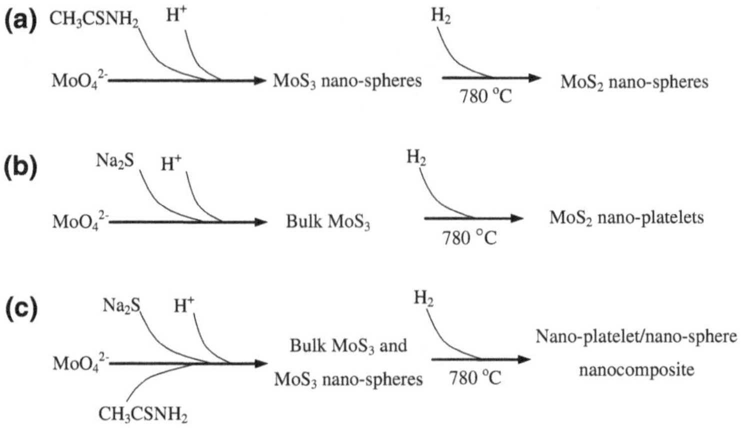

Fig. 8 Schematic of forming nano-MoS_2: **a** nano-spheres, **b** nano-platelets, and **c** MoS_2 nanocomposite with different morphologies (adapted from Ref. [34])

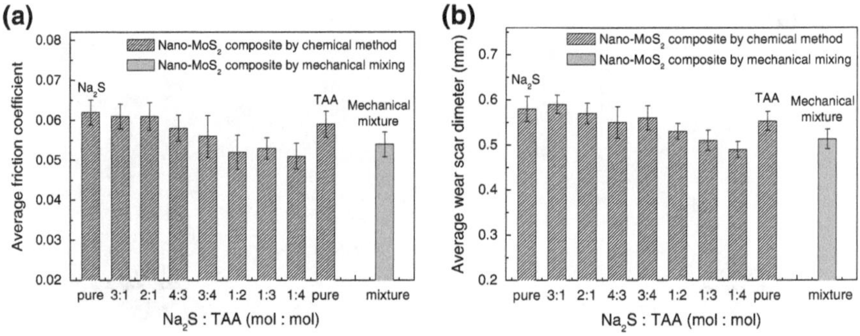

Fig. 9 Variations of friction coefficient (**a**) and wear scar diameter (**b**) lubricated by liquid paraffin with MoS_2 nanocomposites (adapted from Ref. [34])

Figure 9 shows the four-ball tribological properties of the MoS_2 nanocomposite at a rotating speed of 1450 rpm and a constant load of 300 N in liquid paraffin [34]. Figure 9a is the effect of the molar ratio of Na_2S to TAA on the average friction coefficient. As shown in the figure, the LP sample presented the lowest friction coefficient (0.051) at the proportion of 1:4 (Na_2S to TAA). Figure 9b confirms that the variation of AWSD was approximately correlated to the change in friction coefficients. The steel balls lubricated by liquid paraffin with the 1:4 nano-sphere/nano-platelet composite also presented the lowest AWSD (0.49 mm). Compared with the nanocomposites prepared by the mechanically mixing method, the chemically synthesized MoS_2 nanocomposite presented better tribological properties. The chemical method could mix nano-spheres and nano-platelets better than the mechanical one. Thus, the MoS_2 nanocomposite by chemical method showed better tribological properties. However, the difference between the two mixing method is not very obvious. This is mainly because that Na_2S disturbed the forming of nano-sphere.

4.2 MoS_2/Inorganic Compound Nanocomposites

A MoS_3/TiO_2 composite was synthesized by quickly depositing MoS_3 on TiO_2 under a strong acidic solution [35]. Calcining the MoS_3/TiO_2 composite at 450 °C in H_2 led to a MoS_2/TiO_2 nanocomposite. The MoS_2/TiO_2 nanocomposite of 6:5 (wt:wt) was characterized in the literauture. The XRD pattern in Fig. 10a is consistent with that in JCPDS89-4921 belonging to the anatase TiO_2. All diffraction peaks of anatase TiO_2 were still present in the XRD pattern of the MoS_2/TiO_2 nanocomposite (Fig. 10b), indicating that the anatase nano-TiO_2 was not destroyed during the synthesis process. The diffraction peaks of pure nano-MoS_2, reported in Ref. [20], were found in the XRD pattern of the nanocomposite. As shown in Fig. 11a, b, nano-MoS_2 particles were distributed among TiO_2 particles, composed of typical layered structures with an average length of about 15 nm (10–20 nm) and an average thickness of about 5 nm. The nano-MoS_2 particles in the nanocomposite have larger layer distances (~0.66 nm) as compared with pure nano-MoS_2. The findings confirm that the MoS_2/TiO_2 nanocomposite was successfully prepared and provide a new method to synthesize MoS_2-based nanocomposites.

$$3S^{2-} + MoO_4^{2-} + 8H^+ + TiO_2 \longrightarrow MoS_3/TiO_2 \downarrow + 4H_2O.$$

$$MoS_3/TiO_2 + H_2 \xrightarrow{450\,^\circ C/30\ min} nano - MoS_2/TiO_2 + H_2S.$$

The tribological properties of MoS_2/TiO_2 nanocomposite were investigated in liquid paraffin on a four-ball tribometer at 0.556 m/s under 300 N [36]. The MoS_2/TiO_2 nanocomposite was found to be a promising lubricant additive with a better performance than either nano-MoS_2 or nano-TiO_2 alone. Figure 12a provides the variation in average friction coefficient with the mass ratio of MoS_2 to TiO_2 in the nanocomposite. The pure nano-TiO_2 shows the highest friction coefficient and is not an appropriate anti-friction additive in LP. The lowest friction coefficient was observed in the nanocomposite of 2:1 (MoS_2:TiO_2). Figure 12b shows the variation in AWSD with the mass ratio of MoS_2 to TiO_2 in the nanocomposite. The best anti-wear performance was found in the nanocomposite of 4:1. The nanocomposite of 2:1 led to the lowest friction coefficient but an AWSD close to that of pure nano-MoS_2. These mentioned above indicate that forming MoS_2/TiO_2 nanocomposite improved the tribological properties of MoS_2.

Abrasive wear was a main wear factor of steel balls lubricated by LP with the MoS_2/TiO_2 nanocomposite. The nanocomposite, containing higher chemical activity and smaller sizes, could penetrate through the oil film to the contact region.

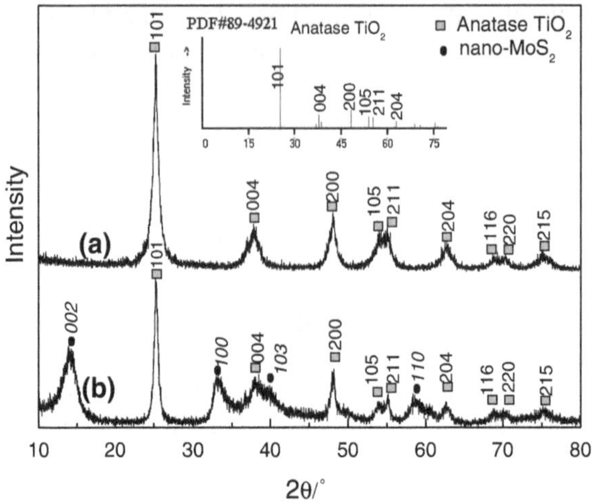

Fig. 10 XRD patterns of: **a** anatase nano-TiO_2 and **b** MoS_2/TiO_2 nanocomposite (adapted from Ref. [35])

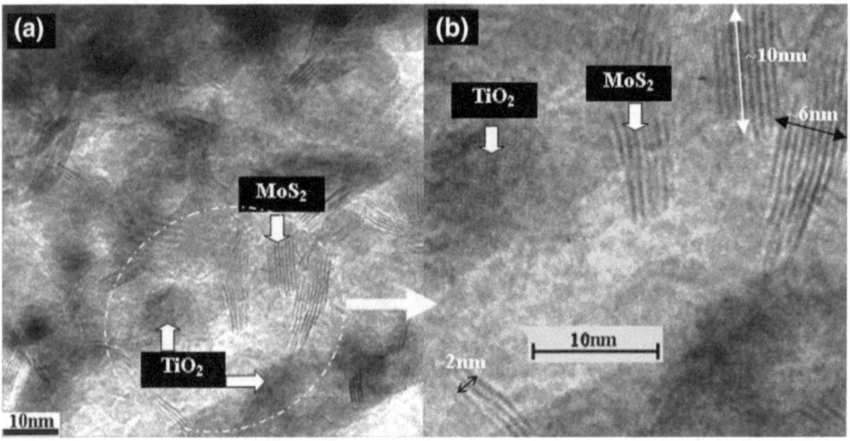

Fig. 11 HRTEM micrographs of the prepared MoS_2/TiO_2 nanocomposite: **a** typical inner region and **b** magnified image of (**a**) (adapted from Ref. [35])

However, the nanoparticles easily agglomerated during the lubrication, leading to inhomogeneous lubrication and asymmetrical furrows (Fig. 13) [23]. Moreover, nano-MoS_2 with the higher chemical activity was more easily reacted with friction pair materials as compared to nano-TiO_2. Thus, the chemical corrosion was also a wear factor of steel balls.

A synergistic effect between nano-TiO_2 and graphite was ascribed to the effective transferring films on friction surfaces and the reinforcing effect of nanoparticles [37]. A transferring film was also found on the steel balls lubricated by the

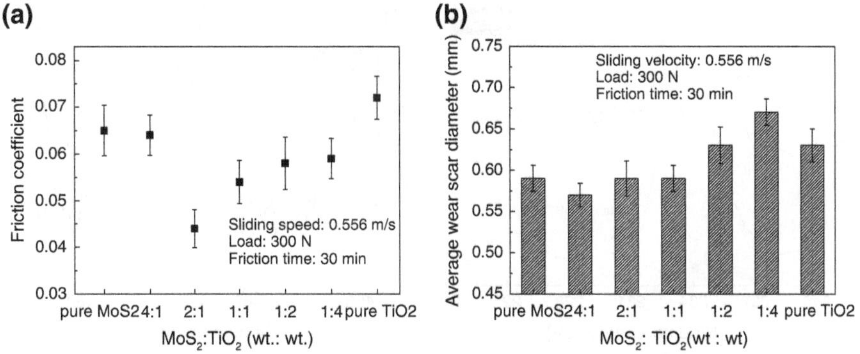

Fig. 12 Tribological properties of MoS_2/TiO_2 nanocomposites: **a** friction coefficient and **b** wear scar diameter (adapted from Ref. [36])

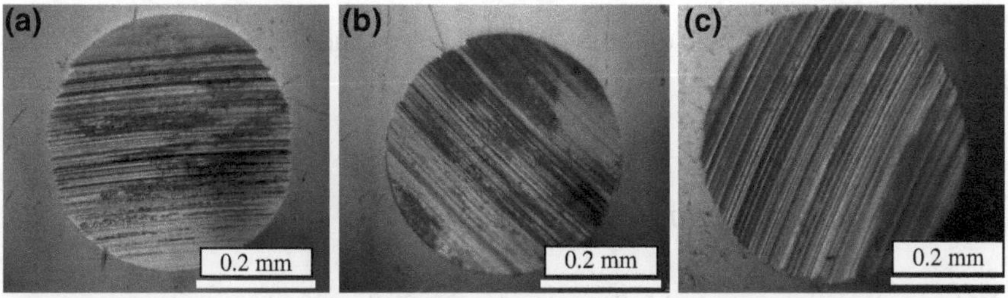

Fig. 13 Optical micrographs of typical wear scars on the bottom balls lubricated by liquid paraffin at 0.556 m/s under 300 N for 30 min with: **a** pure nano-MoS_2, **b** MoS_2/TiO_2 nanocomposite of 4:1, and **c** pure nano-TiO_2 (adapted from Ref. [36])

MoS_2/TiO_2 nanocomposite [36]. The elements Mo and Ti are found in the X-ray photoelectron spectrum (XPS) of the wear scar lubricated by the MoS_2/TiO_2 nanocomposite (Fig. 14) [36]. This implies that MoS_2 and TiO_2 were transferred to the surface of friction pairs from the nanocomposite during friction process. The transfer produced a lubrication film on the steel balls, composed of MoO_3, TiO_2, Fe_2O_3 (or Fe_3O_4), $Fe_2(SO_4)_3$ (or $FeSO_4$), FeS, and carbon-containing compounds after tribochemical reactions.

The excellent lubrication of MoS_2/TiO_2 nanocomposite can also be explained using the effect of nano-TiO_2 on the size and layer distance of nano-MoS_2. Nano-MoS_2 in MoS_2/TiO_2 nanocomposite had smaller thicknesses and larger layer distances as compared to the pure nano-MoS_2. The Large layer distances weakened the Van der Waals force between adjacent MoS_2 molecular layers. Thus, the shearing force needed between these layers decreased. Moreover, the lubrication of the MoS_2/TiO_2 nanocomposite could also be attributed to the micro-cooperation of various nanoparticles with different shapes and lubrication mechanisms [32, 34],

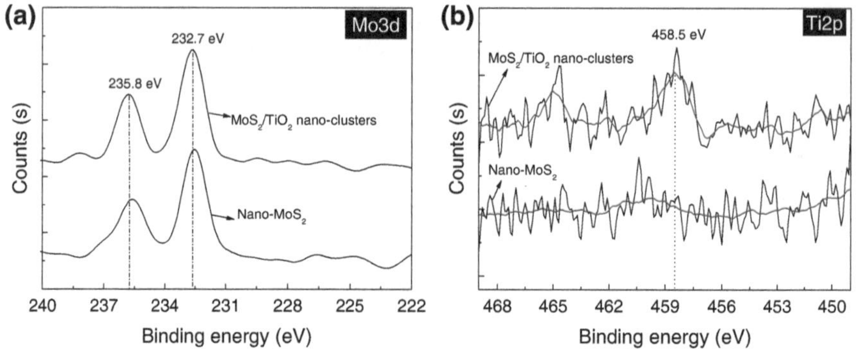

Fig. 14 XPS spectra of wear scars on the top balls lubricated by liquid paraffin at 0.556 m/s under 300 N for 30 min with pure nano-MoS_2 or 2:1 MoS_2/TiO_2 nano-clusters: **a** Mo_{3d} and **b** Ti_{2p} (adapted from Ref. [36])

i.e. the micro-cooperation of MoS_2 nano-platelets and TiO_2 solid nanoparticles during the friction process.

4.3 MoS_2/Polymer Nanocomposites

Mechanically mixing nano-MoS_2 and polymers is the simplest method to prepare MoS_2/polymer nanocomposites for tribological applications. The organic matrix materials mainly included polyoxymethylene (POM) [16, 38–42] and high-density polyethylene (HDPE) [2]. The addition of nano-MoS_2 into polymers had to be done by the heating treatment. It was found that MoS_2 nano-platelet could degrade POM into poisonous formaldehyde in the thermal process (Fig. 15) [16]. Thus, MoS_2 nano-platelet could not be added into POM. Two composites, i.e. MoS_2 micro-platelet/POM and nano-sphere/POM (Fig. 16) [39], were obtained by the mechanical mixing. The nano-sphere/POM revealed better performances in friction reduction and wear resistance as compared to the micro-platelet/POM [40]. Chemical intercalation was an effective chemical method to obtain the MoS_2/POM nanocomposite [3, 42]. The chemical intercalation could disperse MoS_2 better than the mechanical mixing does (Fig. 17) [3]. However, the chmical intercalation destroyed the crystal structure of 2H MoS_2 that is the basis of lubrication. Thus, the intercalation composite did not reveal good lubrication performance (Fig. 18) [40].

HDPE polymer has a more stable structure than POM and the stability of HDPE cannot be affected by nano-platelets at high temperatures [2]. Thus, two nano-MoS_2/HDPE composites, i.e. nano-platelet/HDPE and nano-sphere/HDPE, were prepared by the mechanical mixing (Fig. 19). A fair and interesting comparison was achieved between nano-platelet/HDPE and nano-sphere/HDPE. The

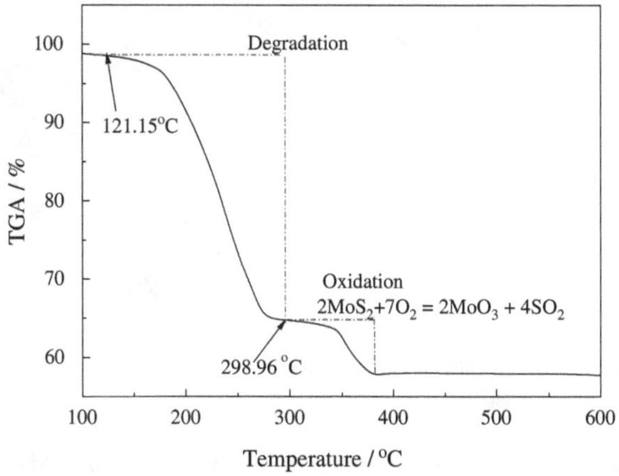

Fig. 15 TGA curve of the mixture of POM powder and MoS_2 nano-platelet (adapted from Ref. [16])

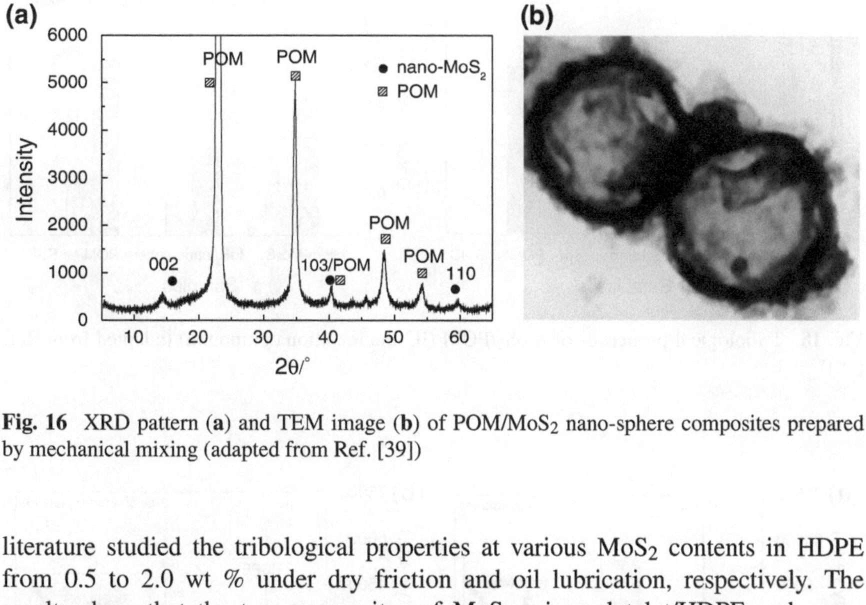

Fig. 16 XRD pattern (**a**) and TEM image (**b**) of POM/MoS_2 nano-sphere composites prepared by mechanical mixing (adapted from Ref. [39])

literature studied the tribological properties at various MoS_2 contents in HDPE from 0.5 to 2.0 wt % under dry friction and oil lubrication, respectively. The results show that the two composites of MoS_2 micro-platelet/HDPE and nano-sphere/HDPE exhibited a similar performance in friction reduction under dry friction. However, the composite with 1.0 wt % MoS_2 nano-platelet showed lower friction coefficients than both micro-platelets/HDPE and nano-spheres/HDPE. The lowest friction coefficient occurred in the composite with 2.0 wt % MoS_2 micro-platelets or nano-spheres (Fig. 20). Under oil lubrication, the nano-sphere/HDPE composite showed the best tribological properties, especially the wear resistance. However, the nano-platelet/HDPE showed no expected tribological properties.

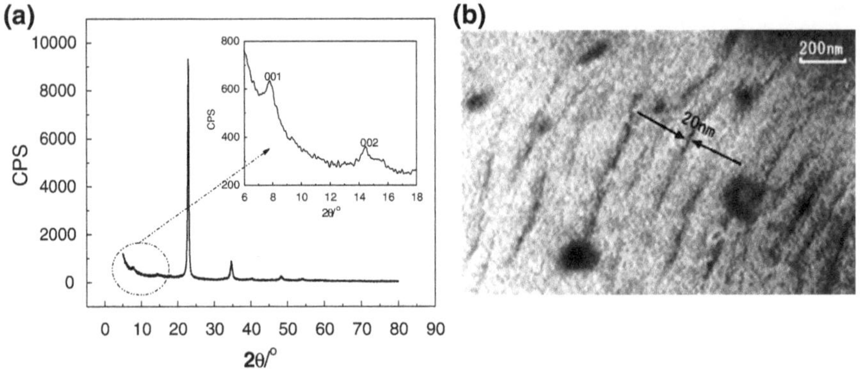

Fig. 17 Intercalation compound of POM/MoS_2: **a** XRD pattern and **b** TEM (adapted from Ref. [3])

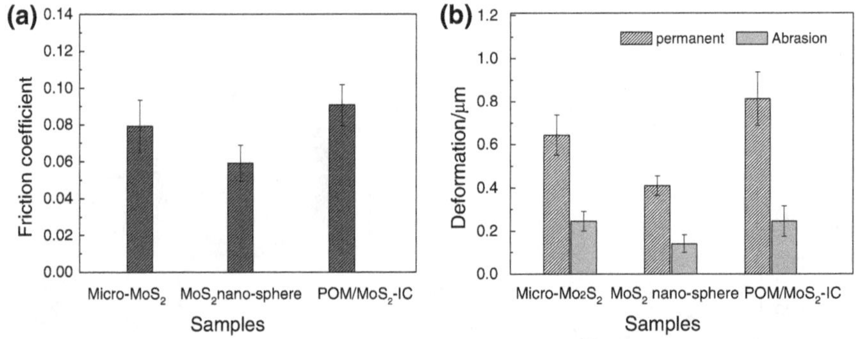

Fig. 18 Tribological properties of MoS_2/POM (IC-intercalation compound) (adapted from Ref. [40])

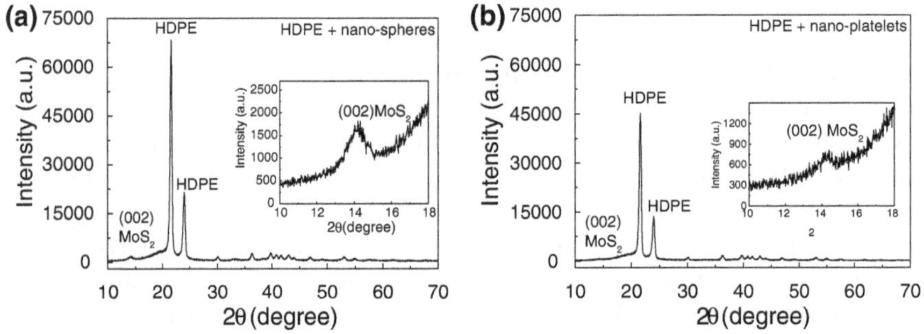

Fig. 19 XRD patterns of: **a** 2.0 % MoS_2 nano-spheres/HDPE and **b** 2.0 % MoS_2 nano-platelets/HDPE (adapted from Ref. [2])

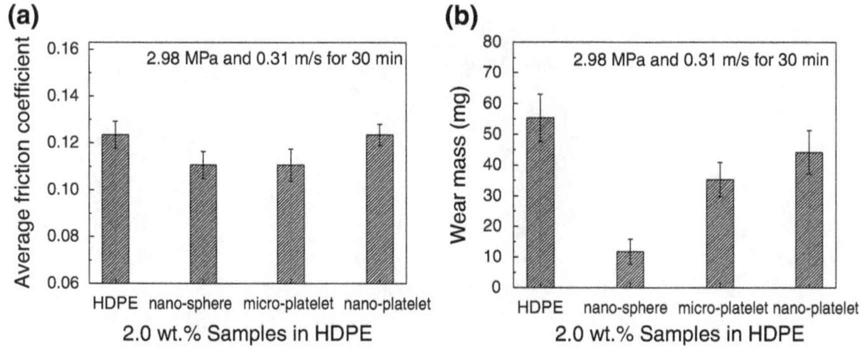

Fig. 20 Tribological properties of MoS_2/HDPE nanocomposites under dry friction: **a** friction coefficient and **b** wear mass (adapted from Ref. [2])

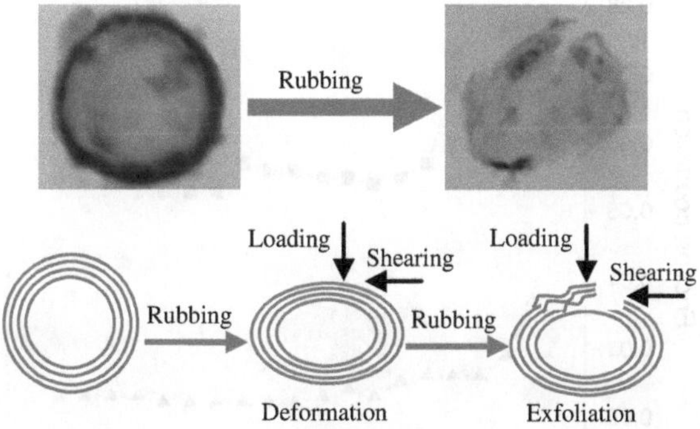

Fig. 21 Schematic diagram of the anti-wear process of MoS_2 nano-sphere (adapted from Ref. [2])

The melting was the main wear mechanism of MoS_2/HDPE composites under dry friction, whereas the abrasive wear became the main wear mechanisms under oil lubrication. The tribological properties of MoS_2/HDPE composites were influenced by their crystallinity and thermo-mechanical behaviors. The addition of nano-sphere into HDPE improved the mechanical behaviors of HDPE, thus leading to better tribological properties. The excellent anti-wear properties of nano-sphere/HDPE composite were attributed to the deformation and exfoliation of the nano-spheres during the friction process (Fig. 21).

4.4 Ni–P–(Nano-MoS_2) Composite Coatings

Ni–P composite coatings with organic or inorganic particles present wide applications in corrosion protection, wear resistance, and friction reduction. Solid lubricants, such as PTFE [43, 44], carbon nanotube [45–47], WS_2 [48], and MoS_2 [49–52], are appropriate additives to modify the Ni–P coatings.

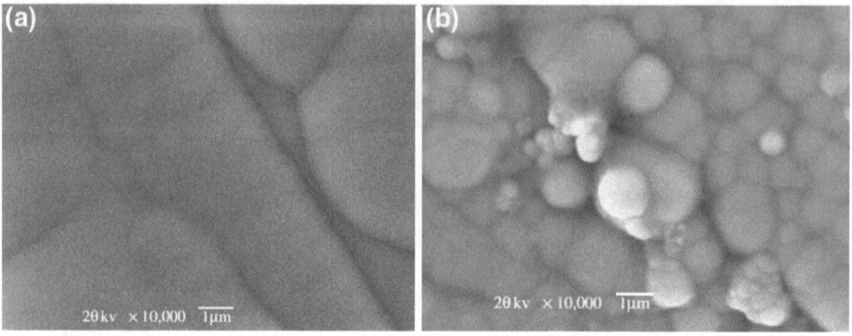

Fig. 22 SEM images of Ni–P electroless coating (**a**) and Ni–P–(nano-MoS_2) electroless coating (**b**) (adapted from Ref. [51])

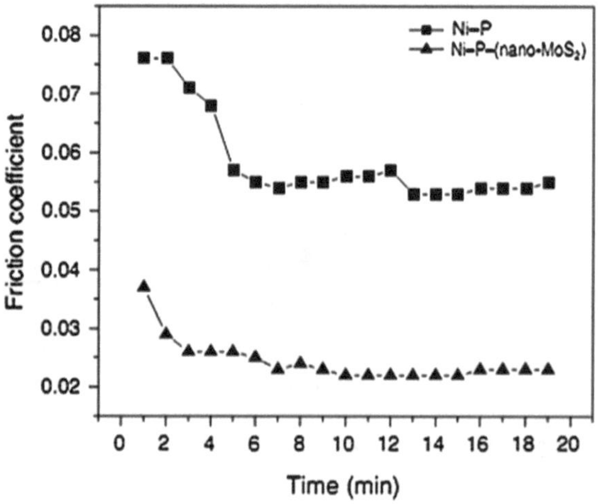

Fig. 23 Friction coefficients of Ni–P and Ni–P–(nano-MoS_2) electroless coatings (adapted from Ref. [51])

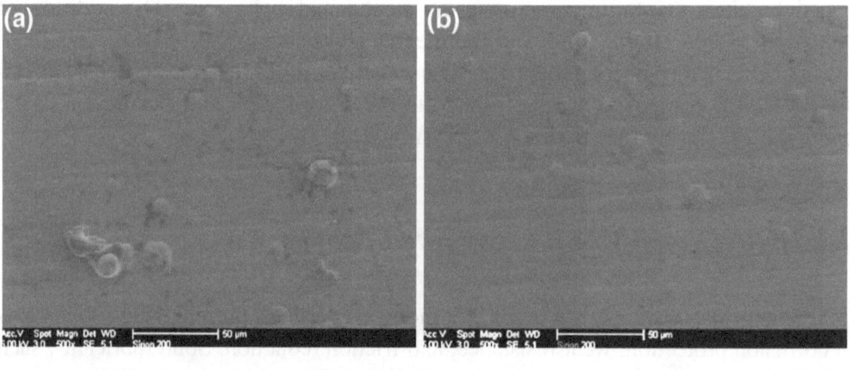

Fig. 24 SEM images of wear surfaces of: Ni–P (**a**) and Ni–P–(nano-MoS_2) (**b**) coatings (adapted from Ref. [51])

Ni–P coatings may be co-deposited with MoS_2 nanoparticles on medium carbon steel substrate by electroless plating [50–52]. Figure 22 shows the SEM images of electroless Ni–P and Ni–P–(nano-MoS_2) composite coatings reported in Ref [51]. The corrosion resistance of the Ni–P–(nano-MoS_2) composite coating was slightly lower than that of the Ni–P coating without MoS_2. As shown in the figure, the cell volume became smaller in the Ni–P–(nano-MoS_2) composite coating as compared to that in the Ni–P coating. It was also found that the nano-MoS_2 particles were around the cell boundary. The Ni–P–(nano-MoS_2) coating showed the super low friction coefficients during the whole rubbing process (Fig. 23). This was attributed to the super lubricity of spherical nano-MoS_2. Moreover, the wear of the Ni–P–(nano-MoS_2) coating was also reduced by the nano-MoS_2 added (Fig. 24).

5 Conclusions

(1) MoS_2-based nanocomposites may be prepared by mechanical mixing, chemical method and electroless coating technology. They usually have better tribological properties than their original materials and play an important role in the lubricating composites.
(2) The chemical method generally reveals advantages over the mechanical one in the preparation of MoS_2 nanocomposites with different morphologies for lubrication applications. However, the chemical intercalation can not improve the tribological properties of MoS_2 nanocomposites, because the intercalation reaction destroys the 2H structure of MoS_2 with better lubrication.
(3) MoS_2/TiO_2 nanocomposite may be prepared by depositing nano-MoS_2 on nano-TiO_2. Nano-MoS_2 and nano-TiO_2 present a positive synergetic effect on the lubrication of the nanocomposite. The sizes of MoS_2 in the nanocomposites are smaller and its layer distances are larger than those of pure nano-MoS_2. Large layer distances weaken the Van der Waals force and small sizes enable MoS_2 to enter the contact region more easily, leading to better anti-friction performance.
(4) Mechanically mixing nano-MoS_2 and polymers, such as POM and HDPE, may produce nano-MoS_2/polymer nanocomposites. MoS_2 nano-sphere in the polymers shows a good lubrication over MoS_2 nano-platelet. The excellent anti-wear properties of nano-spheres are attributed to the deformation and exfoliation of the nano-spheres during the friction process.
(5) Ni–P coatings may be co-deposited with nano-MoS_2 particles on medium carbon steel substrate by electroless plating. The co-deposited nano-MoS_2 significantly improves the friction reduction of Ni–P coating.

Acknowledgments This work was financially supported by the National Natural Science Foundation of China (Grant No. 50905054 and 51275143), China Postdoctoral Science Foundation funded Project (Grant No. 2011M500110), Foundation of State Key Laboratory of Solid Lubrication (Grant No. 0907), and Talents Foundation of Hefei University (Grant No. 12RC03).

References

1. Benavente E, Santa Ana MA, Mendizábal F, González G (2002) Intercalation chemistry of molybdenum disulfide. Coord Chem Rev 224:87–109
2. Hu KH, Hu XG, Wang J, Xu YF, Han CL (2012) Tribological properties of MoS_2 with different morphologies in high-density polyethylene. Tribol Lett 47:79–90
3. Wang J, Hu KH, Xu YF, Hu XG (2008) Structural, thermal, and tribological properties of intercalated polyoxymethylene/molybdenum disulfide nanocomposites. J Appl Polym Sci 110:91–96
4. Zhang Z, Liu W, Xue Q, Zen J (1998) Current state of tribological application and research of Mo compounds as lubricating materials. Tribology 18(4):377–382
5. Dunckle CG, Aggleton M, Glassman J, Taborek P (2011) Friction of molybdenum disulfide–titanium films under cryogenic vacuum conditions. Tribol Int 44:1819–1826
6. Kalin M, Kogovšek J, Remškar M (2012) Mechanisms and improvements in the friction and wear behavior using MoS_2 nanotubes as potential oil additives. Wear 280–281:36–45
7. Luo J, Zhu MH, Wang YD, Zheng JF, Mo JL (2011) Study on rotational fretting wear of bonded MoS_2 solid lubricant coating prepared on medium carbon steel. Tribol Int 44:1565–1570
8. Feldman Y, Wasserman E, Srolovitz DJ, Tenne R (1995) High rate, gas phase growth of MoS_2 nested inorganic fullerenes and nanotubes. Science 267:222–225
9. Nath M, Govindaraj A, Rao CNR (2001) Simple synthesis of MoS_2 and WS_2 nanotubes. Adv Mater 13:283–286
10. Li XL, Li YD (2004) MoS_2 nanostructures: synthesis and electrochemical Mg^{2+} intercalation. J. Phys Chem B 108:13893–13900
11. Hu KH, Hu XG (2009) Formation, exfoliation and restacking of MoS_2 nanostructures. Mater Sci Technol 25:407–414
12. Hu KH, Wang YR, Hu XG, Wo HZ (2007) Preparation and characterisation of ball-like MoS_2 nanoparticles. Mater Sci Technol 23:242–246
13. Lavayen V, Mirabal N, O'Dwyer C, Santa Ana MA, Benavente E, Sotomayor Torres CM, González G (2007) The formation of nanotubes and nanocoils of molybdenum disulphide. Appl Surf Sci 253:5185–5190
14. Cizaire L, Vacher B, Mogne TL, Martin JM, Rapoport L, Margolin A, Tenne R (2002) Mechanisms of ultra-low friction by hollow inorganic fullerene-like MoS_2 nanoparticles. Surf Coat Technol 160:282–287
15. Rapoport L, Feldman Y, Homyonfer M, Cohen H, Sloan J, Hutchison JL, Tenne R (1999) Inorganic fullerene-like material as additives to lubricants: structure–function relationship. Wear 225–229:975–982
16. Hu KH, Hu XG, Sun XJ (2010) Morphological effect of MoS_2 nanoparticles on catalytic oxidation and vacuum lubrication. Appl Surf Sci 256:2517–2523
17. Rapoport L, Moshkovich A, Perfilyev V, Laikhtman A, Lapsker I, Yadgarov L, Rosentsveig R, Tenne R (2012) High lubricity of re-doped fullerene-like MoS_2 nanoparticles. Tribol Lett 45:257–264
18. Zou TZ, Tu JP, Huang HD, Lai DM, Zhang LL, He DN (2006) Preparation and tribological properties of inorganic fullerene-like MoS_2. Adv Eng Mater 8:289–293
19. Hu KH, Liu M, Wang QJ, Xu YF, Schraube S, Hu XG (2009) Tribological properties of molybdenum disulfide nanosheets by monolayer restacking process as additive in liquid paraffin. Tribol Int 42:33–39
20. Chhowalla M, Amaratunga GAJ (2000) Thin films of fullerene-like MoS_2 nanoparticles with ultra-low friction and wear. Nature 407:164–167

21. Rosentsveig R, Gorodnev A, Feuerstein N, Friedman H, Zak A, Fleischer N, Tannous J, Dassenoy F, Tenne R (2009) Fullerene-like MoS_2 nanoparticles and their tribological behavior. Tribol Lett 36:175–182
22. Rapoport L, Fleischer N, Tenne R (2005) Applications of WS_2 (MoS_2) inorganic nanotubes and fullerene-like nanoparticles for solid lubrication and for structural nanocomposites. J Mater Chem 15:1782–1788
23. Hu KH, Hu XG, Xu YF, Huang F, Liu JS (2010) The effect of morphology on the tribological properties of MoS_2 in liquid paraffin. Tribol Lett 40:155–165
24. Tannous J, Dassenoy F, Lahouij I, Le Mogne T, Vacher B, Bruhács A, Tremel W (2011) Understanding the tribochemical mechanisms of IF-MoS_2 nanoparticles under boundary lubrication. Tribol Lett 41:55–64
25. Lahouij I, Dassenoy F, Vacher B, Martin JM (2012) Real time TEM imaging of compression and shear of single fullerene-like MoS_2 nanoparticle. Tribol Lett 45:131–141
26. Lahouij I, Dassenoy F, de Knoop L, Martin JM, Vacher B (2011) In situ TEM observation of the behavior of an individual fullerene-Like MoS_2 nanoparticle in a dynamic contact. Tribol Lett 42:133–140
27. Daage M, Chianelli RR (1994) Structure-function relations in molybdenum sulfide catalysts: the "rim-edge" model. J Catal 149:414–427
28. Wo HZ, Hu KH, Hu XG (2004) Tribological properties of MoS_2 nanoparticles as additive in a machine oil. Tribology 24:33–37
29. Wang TM, Shao X, Wang QH, Liu WM (2005) Preparation and tribological behavior of polyimide MoS_2 intercalation composite. Tribology 25:322–327
30. Joly-Pottuz L, Martin JM, Dassenoy F, Belin M, Montagnac R, Reynard B (2006) Pressure-induced exfoliation of inorganic fullerene-like WS_2 particles in a Hertzian contact. J Appl Phys 99:023524–023528
31. Tevet O, Goldbart O, Cohen SR, Rosentsveig R, Popovitz-Biro R, Wagner HD, Tenne R (2010) Nanocompression of individual multilayered polyhedral nanoparticles. Nanotechnology 21:365705–365711
32. Hu KH, Cai YK, Hu XG, Xu YF (2010) Synergistic lubrication of MoS_2 particles with different morphologies in liquid paraffin. Ind Lubr Tribol, Ref: ilt-10-2010-0075, accepted manuscript
33. Hu KH, Hu XG, Jiang P (2010) Large-scale and morphology-controlled synthesis of nano-sized molybdenum disulphide particles by different sulphur sources. Int J Mater Prod Technol 39:378–387
34. Hu KH, Cai YK, Hu XG, Xu YF (2011) Synthesis and tribological properties of MoS_2 composite nanoparticles with different morphologies. Surf Eng 27:544–550
35. Hu KH, Hu XG, Xu YF, Sun JD (2010) Synthesis of nano-MoS_2/TiO_2 composite and its catalytic degradation effect on methyl orange. J Mater Sci 45:2640–2648
36. Hu KH, Huang F, Hu XG, Xu YF, Zhou YQ (2011) Synergistic effect of nano-MoS_2 and anatase nano-TiO_2 on the lubrication properties of MoS_2/TiO_2 nano-clusters. Tribol Lett 43:77–87
37. Xian G, Walter R, Haupert F (2006) A synergistic effect of nano-TiO_2 and graphite on the tribological performance of epoxy matrix composites. J Appl Polym Sci 102:2391–2400
38. Hu KH, Wang J, Schraube S, Xu YF, Hu XG, Stengler R (2009) Tribological properties of MoS_2 nano-balls as filler in plastic layer of three-layer self-lubrication bearing materials. Wear 266:1198–1207
39. Hu KH, Sun XJ, Xu YF, Salomon M, Hu XG (2009) Tribological properties of POM-based self-lubrication composites with MoS_2 in vacuum. J Hefei Univ Technol 32:615–619
40. Hu KH, Schraube S, Xu YF, Hu XG, Stengler R (2010) Micro-tribological behavior of polyacetal-based self-Lubrication composite materials modified with MoS_2. Tribology 30:38–45
41. Hu X, Hu K, Sun X, Xu Y (2010) Space tribological properties of MoS_2-based self-lubrication composite materials. Spacecraft Environ Eng 27(1):50–53
42. Hu KH, Xu YF, Wang J, Hu XG, Schraube S, Stengler R (2007) Tribological behavior of self-lubrication bearing materials of steel-copper-polyoxymethylene containing MoS_2-IC nanoparticles. ASME/STLE 2007 international joint tribology conference (IJTC2007), 2007, San Diego, California, USA, pp 787–789

43. Ramalho A, Miranda JC (2005) Friction and wear of electroless NiP and NiP + PTFE coatings. Wear 259:828–834
44. Rossi S, Chini F, Straffelini G, Bonora PL, Stampali A (2003) Corrosion protection of electroless nickel/PTFE, phos-phate/MoS_2 and bronze/PTFE coatings applied to improvethe wear resistance of carbon steel. Surf Coat Technol 173:235–242
45. Chen WX, Tu JP, Xu ZD, Chen WL, Zhang XB, Cheng DH (2003) Tribological properties of Ni-P-multi-walled carbon nanotubes electroless composite coating. Mater Lett 57:1256–1260
46. Wang LY, Tu JP, Chen WX, Wang YC, Liu XK, Olk C, Cheng DH, Zhang XB (2003) Friction and wear behavior of electroless Ni-based CNT composite coatings. Wear 254:1289–1293
47. Chen XH, Chen CS, Xiao HN, Cheng FQ, Zhang G, Yi GJ (2005) Corrosion behavior of carbon nanotubes-Ni composite coating. Surf Coat Technol 191:351–356
48. Chen WX, Tu JP, Ma XC, Xu ZD, Chen WL, Wang JG, Cheng DH, Xia JB, Gan HY, Jin YX, Tenne R, Rosenstveig R (2002) Preparation and tribological properties of Ni-P electroless composite coating containing inorganic fullerene-Like WS_2 nanomaterials. Acta Chim Sin 60:1722–1726
49. Zou TZ, Tu JP, Zhang SC, Chen LM, Wang Q, Zhang LL, He DN (2006) Friction and wear properties of electroless Ni-P- (IF-MoS_2) composite coatings in humid air and vacuum. Mater Sci Eng, A 426:162–168
50. Hu XG, Cai WJ, Xu YF, Wan JC, Sun XJ (2009) Electroless Ni–P–(nano-MoS_2) composite coatings and their corrosion properties. Surf Eng 25:361–366
51. Hu X, Jiang P, Wan J, Xu Y, Sun X (2009) Study of corrosion and friction reduction of electroless Ni–P coating with molybdenum disulfide nanoparticles. J Coat Technol Res 6:275–281
52. Hu XG, Cai WJ, Wan JC, Xu YF, Sun XJ. Study of corrosion performance of electroless Ni-P coating with molybdenum disulfide nanoparticles. Key Eng Mater 373–374:256

Friction and Wear of Al_2O_3-Based Composites with Dispersed and Agglomerated Nanoparticles

Jinjun Lu, Jian Shang, Junhu Meng and Tao Wang

Abstract In recent years, Al_2O_3-based nanocomposite which is composed of micro-size Al_2O_3 matrix and a second phase nanoparticles (e.g. SiC, Ni) as the reinforcement phase have received considerable attention and come a long way mainly because of their good mechanical property and resistance to abrasive wear and erosive wear. Up to now, practice on tribologcial design and testing of high temperature self-lubricating Al_2O_3-based nanocomposites is an exciting field to be explored. In this chapter, the principle for fabrication of Al_2O_3-based nanocomposite for better mechanical property than that of monolithic Al_2O_3 ceramic is briefly introduced on basis of literature review. In other words, the microstructure, and mechanical property of Al_2O_3-SiC, Al_2O_3-Ni nanocomposites are briefly introduced and discussed. This chapter focuses on the design and tribological property of high temperature self-lubricating Al_2O_3-based nanocomposites. As the first step, the tribological consideration of Al_2O_3-SiC nanocomposite is discussed according to Todd's work which gives useful information on the pullouts of grains during grinding and polishing. In addition, a concept for designing high temperature self-lubricating Al_2O_3-based nanocomposite with dispersed and agglomerated ceramic nanoparticles is proposed and discussed using Al_2O_3-TiCN composite as an example. Al_2O_3-5 wt % TiCN and Al_2O_3-10 wt % TiCN composites had high hardness and good tribological property at room temperature in air. Al_2O_3-TiCN composites containing 10 wt % and 20 wt % TiCN nanoparticles in sliding against Ni-Cr alloy are self-lubricating at 500 °C.

J. Lu (✉) · J. Shang · J. Meng · T. Wang
State Key Laboratory of Solid Lubrication, Lanzhou Institute of Chemical Physics,
Chinese Academy of Sciences, 730000 Lanzhou, China
e-mail: jjlu@licp.cas.cn

J. Paulo Davim (ed.), *Tribology of Nanocomposites*, Materials Forming,
Machining and Tribology, DOI: 10.1007/978-3-642-33882-3_4,

1 Introduction

Alumina (Al_2O_3) is one of the five important classes of advanced structural ceramics. In addition, ceramics based on Al_2O_3 have been used in commercial applications for many years because of their availability and low cost [1]. There are many efforts to improve the mechanical and tribological property of Al_2O_3 ceramic by adding a second phase particles. The particles are mostly micro-size particles.

In the 1990s, the concept of Al_2O_3-based nanocomposite was proposed and two representatives of this kind of nanocomposites, i.e. Al_2O_3-SiC and Al_2O_3-Ni, have been successfully fabricated. The mechanical property of this kind of nanocomposite is very appealing. For specification, some of the reported bending strengths of the two nanocomposites are as high as 1 GPa which is much higher than that of monolithic Al_2O_3 ceramic [2, 3]. This result has driven many researchers to reveal the strengthening mechanism of SiC and Ni nanoparticles. Despite their very high bending strength, the fracture toughness of Al_2O_3-SiC and Al_2O_3-Ni nanocomposites is not high enough to make them a tough material. The discussion will be forwarded later in Sect. 1.1.2.

In this section, the principle for fabrication of Al_2O_3-based nanocomposite for better mechanical property than that of monolithic Al_2O_3 ceramic is introduced on basis of literature review. To specify, the microstructure, mechanical property and tribological property of Al_2O_3-based nanocomposite will be introduced and discussed.

1.1 Microstructure, Mechanical and Tribological Property of Al_2O_3-Based Nanocomposite

1.1.1 Microstructure

The main weakness of monolithic Al_2O_3 ceramic as a structural material is its intrinsic brittleness. To make Al_2O_3 ceramic tougher, a second phase in form of particle, whisker, and fiber is often added. The micro-size particle, whisker, and fiber can indeed increase the fracture toughness of Al_2O_3 ceramic [2]; however, the sacrifice of strength is the drawback of this kind of toughening mechanism.

In the 1990s, a new concept is proposed to solve the above problem. The microstructure of Al_2O_3-based nanocomposite can be classified into two groups: micro-size Al_2O_3 matrix with nano-size particle or whisker (group I) and nano-size Al_2O_3 matrix with nano-size particle or whisker (group II). The group I can be further divided into three subgroups according to the location of nano-size particle or whisker. The three subgroups are inter-type, intra-type and inter/intra-type. Up to now, Al_2O_3-based nanocomposite in the published papers belongs to group I. In the following section, unless otherwise stated, Al_2O_3-based nanocomposite refers to Al_2O_3-based nanocomposite of group I. Al_2O_3-based nanocomposite of group II is expected to have several attractive functions, e.g. superplasticity and good machinability but is still an myth now. The type of Al_2O_3-based nanocomposite is summarized in Table 1.

Table 1 Type of Al_2O_3-based nanocomposite

Group	Sub-Group	Description
I. Micro-nano type	Inter-type Intra-type Inter/intra-type	Micro-size Al_2O_3 matrix, submicro-size or nano size second phase particles
II. Nano-nano type		Nano-size Al_2O_3 matrix, nano-size second phase particles

The second phase particles for Al_2O_3-based nanocomposite in Table 1 are generally required to be well dispersed rather than agglomerated. However, the agglomerated nanoparticles are sometimes useful from viewpoint of tribology and will be discussed later.

1.1.2 Mechanical Property

The addition of nano-size particle or whisker in Al_2O_3-based nanocomposite has at least two beneficial effects on mechanical property. One is the fine grain Al_2O_3 matrix and the other is the pinning effect of nano-size particle or whisker.

The initial aim for preparing Al_2O_3-based nanocomposite is to increase the fracture toughness. According to the published data, the increment in fracture toughness is very limited. On the other hand, it is very interesting that the three-point bending strength of Al_2O_3-based nanocomposite is high up to 1 GPa. The high bending strength indicates that Al_2O_3-based nanocomposite is capable of tolerating high tensile stress while the limited improvement on fracture toughness indicates Al_2O_3-based nanocomposite is still a brittle material.

It is convincing to elucidate the high bending strength of Al_2O_3-based nanocomposite from Todd's work [4–6]. As it will be shown in the next section, the area fraction of pullout of grains and pullout diameter on the surface of Al_2O_3-based nanocomposite are lower than that of monolithic Al_2O_3 ceramic. As a consequence, the number and size of the defects on a beam made of Al_2O_3-based nanocomposite for a three-point bending test are lower than that on a beam made of monolithic Al_2O_3 ceramic. It is the defect (pullout of grain) that determines the difference in bending strength of Al_2O_3-based nanocomposite and monolithic Al_2O_3 ceramic. The contribution of reducing pullout of grains to the fracture toughness is limited.

1.1.3 Tribological Property

Findings indicate that Al_2O_3-based nanocomposite exhibits better tribological performance than that of monolithic Al_2O_3 ceramic [7, 8]. Of these tribological investigations, the tribological property of Al_2O_3-based nanocomposite is mainly on friction and wear dominated by fracture and pullout of grains, e.g. abrasive wear and erosive wear.

The abrasive wear of Al_2O_3-SiC nanocomposite is systematically investigated by Todd and his main contribution to this field is the discovery of quantity and type of pullout of grains. He uses two parameters, area fraction of pullout and pullout diameter, for comparison of the wears of Al_2O_3-SiC nanocomposite and monolithic Al_2O_3 ceramic. The area fraction of pullout and pullout diameter for Al_2O_3-SiC nanocomposite are lower than that of monolithic Al_2O_3 ceramic. The fraction of transgranular and intergranular pullout is also useful for understanding the wear mechanism. The results of abrasive wear of Al_2O_3-SiC nanocomposite is very helpful for the understanding of high three-point bending strength mentioned in Sect. 1.1.2. The surface defects on the beam of Al_2O_3-SiC nanocomposite for a three-point bending test are mainly pullout of grain and the number of surface defects can be greatly reduced during the grinding and polishing.

Fine grain and less number of surface defect are also very useful for good wear resistance for erosive wear. In this sense, the wear resistance of Al_2O_3-SiC nanocomposite to erosive wear is good.

1.2 Tribological Consideration of High Temperature Self-Lubricating Al_2O_3-Based Nanocomposite

Up to now, the tribological properties of Al_2O_3-SiC and Al_2O_3-Ni nanocomposites including sliding friction and wear as well as abrasive wear have been evaluated. And some results are quite good indeed.

However, as is known to the readers, the basis principle for designing these Al_2O_3-based nanocomposite is based on the viewpoint of improving the mechanical strength, not the tribological property. The principles for fabrication of a structural material and a tribo-material are common and uncommon in many aspects. There are two examples. The first one is about the common aspect. Fine grain Al_2O_3 matrix is preferred for improving mechanical and tribological properties. The second one is about the uncommon aspect. The second phase particles in Al_2O_3-based nanocomposite can be well dispersed and agglomerated for good tribological behavior but only well dispersed for good mechanical property. That is to say, an optimized microstructure of an Al_2O_3-based nanocomposite for high mechanical strength might not be an optimized one for low friction and/or high wear resistance. Therefore, it is urgent to propose material design from the viewpoint of tribology.

In this chapter, we propose a concept which is derived from the concept in Table 1, to design high temperature self-lubricating Al_2O_3-based nanocomposite. The proposed high temperature self-lubricating Al_2O_3-based nanocomposite is composed of a micro-size Al_2O_3 matrix and a second phase nanoparticle. The second phase nanoparticle should be served as both strengthening component and tribological component. In this sense, some solid lubricants, e.g. MoS_2, graphite, CuO, PbO, are not qualified as strengthening component. Ni, Cu and Mo are a good choice because they can improve the bending strength. In addition, the oxides of Ni, Cu and Mo are lubricious at high temperature. The two points can make Al_2O_3-Ni (Cu, W) nanocomposite as high temperature self-lubricating materials. Another group of the second phase nanoparticle contains hard carbides (e.g. SiC, TiC, WC, TiCN) and borides (e.g. TiB_2). These carbides and borides have high melting point and hardness and can strengthen Al_2O_3 ceramic. The nanoparticles of carbides and borides can be readily oxidized at high temperature and the oxides are lubricious. Besides dispersed nanoparticles, agglomerated nanoparticles might be helpful for the formation of continuous lubricious oxide film on the worn surface.

In summary, the basic points of the concept are: (1) Fine grain Al_2O_3 matrix with typical grain size of less than 3 μm. (2) The second phase nanoparticles can strengthen Al_2O_3 ceramic. (3) Lubrication is provided by lubricious oxide generated by tribo-oxidation of nanoparticles at high temperature. (4) Agglomerated nanoparticles give birth to continuous lubricious oxide film on the worn surface.

The tribological behaviors of a Ni-based alloy/Al_2O_3 tribo-pair at elevated temperatures are of theoretical importance. The dominate wear mechanism of a metal/Al_2O_3 tribo-pair at high temperature is adhesive wear which is evident by severe material transfer from alloy to Al_2O_3. Some oxides (e.g. CuO) and tribo-oxides (TiO_2 from TiCN, TiC) can eliminate or prevent the material transfer from Ni-based alloy to Al_2O_3 and reducing the friction coefficient. TiCN has a high hardness comparable to SiC and can be used as a second phase for improving the mechanical strength of Al_2O_3. The tribo-oxide of TiCN at high speed and elevated temperatures can provide effective lubrication on the frictional surface of Al_2O_3. This means TiCN can be used as both strengthening component and tribological component for Al_2O_3. From the viewpoint of sintering, the sintering temperature of TiCN is higher than that of Al_2O_3 and this enables the dispersed and agglomerated TiCN nanoparticles in the Al_2O_3 matrix. In this connection, Al_2O_3-TiCN composite with dispersed and agglomerated nanoparticles is used as an example of this chapter. In the following sections, the preparation, microstructure, mechanical property and tribological property of Al_2O_3-TiCN composite will be introduced.

2 Preparation, Microstructure and Mechanical Property of Al_2O_3-TiCN Composite

2.1 Material

2.1.1 Preparation

Al_2O_3 and $Ti(C_{0.3}N_{0.7})$ powders to fabricate Al_2O_3-TiCN composite were commercially available from Nanjing Emperor Nano Material Co., Ltd. and Shijiazhuang Huatai Nanoceramic Factory. The average particle sizes of Al_2O_3 and $Ti(C_{0.3}N_{0.7})$ powders are 0.5 μm and 50 nm, respectively.

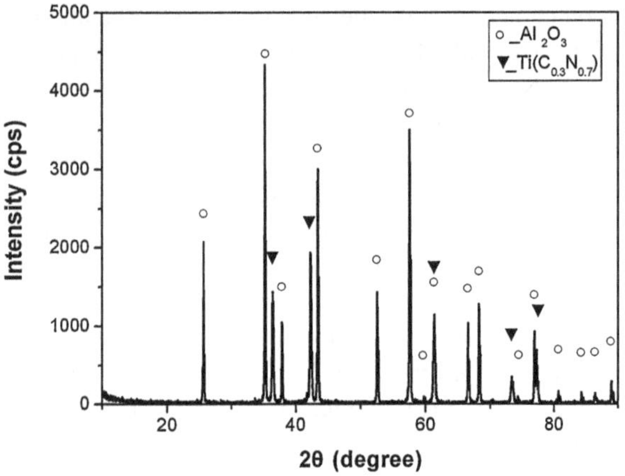

Fig. 1 The XRD pattern of A10T composite

Al_2O_3 and $Ti(C_{0.3}N_{0.7})$ powders were mechanically mixed in a HF-G7-R75S2 Planetary Grinding Machine (Nanjing NanDa Instrument Plant, China) using agate balls and alcohol as grinding media. The weight ratio of ball and powder was 10:1. Composite powders with mass fractions of 5, 10, 20, 30, and 40 % for $Ti(C_{0.3}N_{0.7})$ were prepared to fabricate corresponding Al_2O_3-TiCN composites. In the following section, A5T, A10T, A20T, A30T and A40T are abbreviations for Al_2O_3-TiCN composites with mass fractions of 5, 10, 20, 30, and 40 % of $Ti(C_{0.3}N_{0.7})$.

One-step sintering was used to fabricate Al_2O_3-TiCN composite. The mixed Al_2O_3-$Ti(C_{0.3}N_{0.7})$ powders were hot pressed in a graphite die using a ZT-63-20Y vacuum hot-pressing sintering furnace (manufactured by Shanghai Chen Hua Electric Furnace Co. Ltd.). The hot pressing was conducted at a sintering temperature of 1,400 °C and a pressure of 30 MPa for 120 min in Ar gas. The sintering parameters was optimized for monolithic Al_2O_3 ceramic, not for Al_2O_3-TiCN composites in this chapter. Moreover, a sintering temperature of 1,400 °C is not high enough for sintering and apparent grain growth of TiCN particles. As a result, it is important that TiCN particles remain as nanoparticles in the composite. In addition, the agglomerated TiCN nanoparticles in Al_2O_3-TiCN composite can be achieved by adding adequate amount of TiCN particles.

2.1.2 Microstructure and Mechanical Property

The microstructure of Al_2O_3-TiCN composites was investigated using X-ray diffraction (XRD) to determine the phase composition and scanning electron microscopy (SEM) to determine the grain sizes of Al_2O_3 and TiCN and distribution of TiCN particles.

XRD results show that all the Al_2O_3-TiCN composites were composed of two phases, i.e. α-Al_2O_3 and TiCN; no chemical reaction between Al_2O_3 and TiCN and no decomposition of TiCN occurred, see Fig. 1.

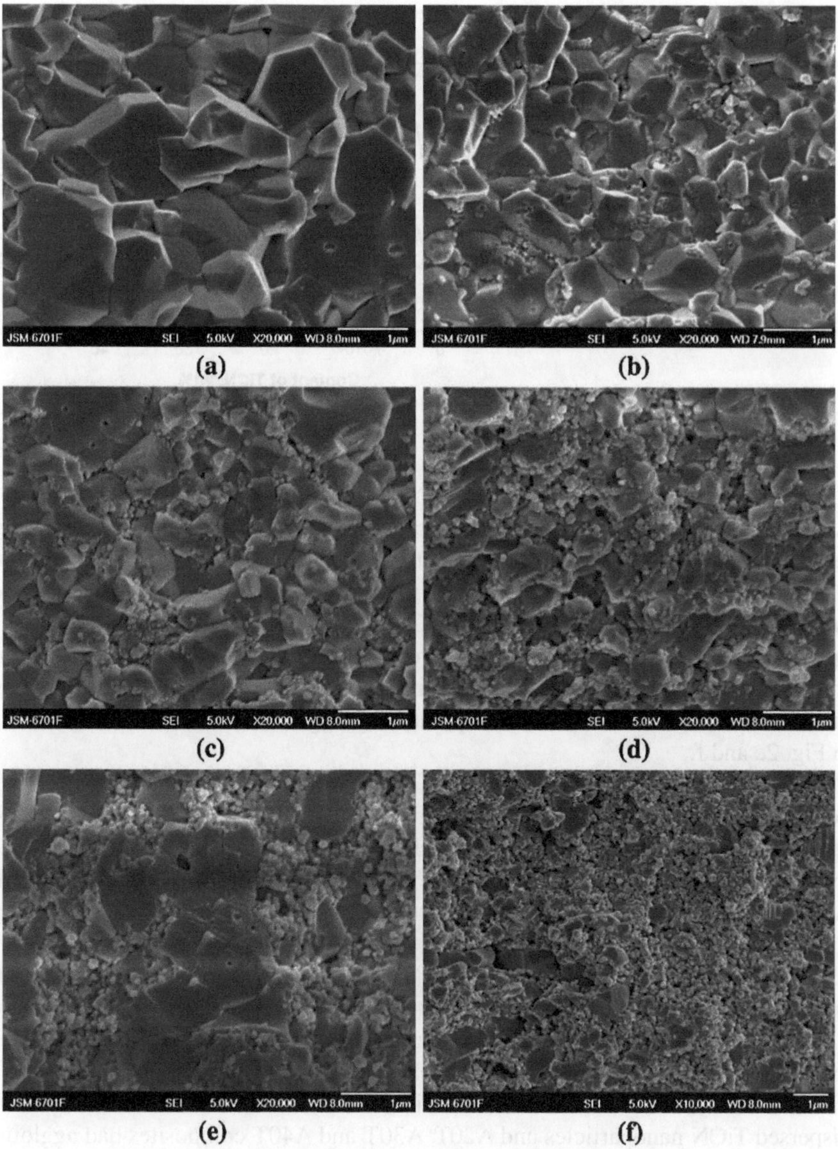

Fig. 2 SEM micrographs of fractured surface of monolithic Al_2O_3 ceramic and Al_2O_3-TiCN composites. The mass fractions of TiCN are **a** 0 %, **b** 5 %, **c** 10 %, **d** 20 %, **e** 30 %, **f** 40 %

Monolithic Al_2O_3 ceramic in Fig. 2a had a dense and fine grain microstructure. The average grain size d50 was 3 μm. As seen in Fig. 2, by varying the addition amount of TiCN nanoparticles, the Al_2O_3-TiCN composites exhibited evolution of microstructures in some aspects, i.e. the grain size and distribution of Al_2O_3; distribution of TiCN nanoparticles; agglomeration of TiCN nanoparticles.

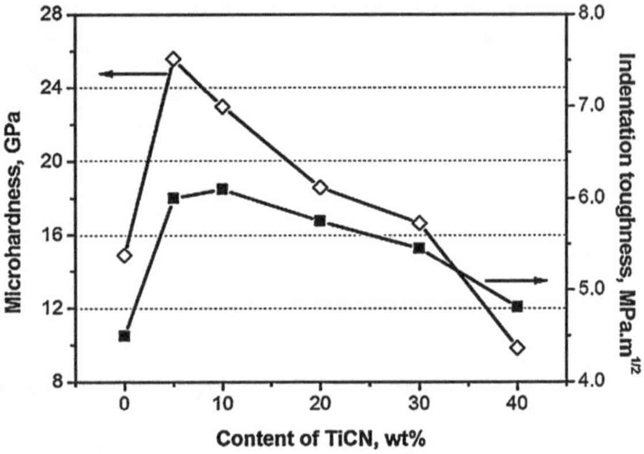

Fig. 3 Microhardness and indentation toughness of the Al_2O_3-TiCN composites

In general, the incorporation of TiCN nanoparticles inhibited the growth of Al_2O_3 grain and thereby reduced the grain size of Al_2O_3 at the same sintering temperature. The grain sizes of Al_2O_3 in Fig. 2b to 2f were smaller than that in Fig. 2a. The addition of 5 wt % TiCN nanoparticles can effectively reduce the grain size of Al_2O_3 to ca. 1 μm, see Fig. 2b. Higher addition amount of TiCN nanoparticles, i.e. 10 wt % and more, can reduce the grain size of Al_2O_3; however, some large Al_2O_3 grains can also be found, see Fig. 3e. Al_2O_3 was considered as the matrix in Fig. 2b to d but was separated by agglomerated TiCN nanoparticles in Fig. 2e and f.

For A5T composite, the distribution of TiCN nanoparticles in Al_2O_3 matrix was not even with very small fraction of agglomerated TiCN nanoparticles; and both inter-type and intra-type TiCN nanoparticles can be found. In this chapter, the relative amount of inter-type and intra-type TiCN nanoparticles was not determined. A10T composite had a similar microstructure as that of A5T composite comparing Figs. 2b and 2c.

The degree of agglomeration of TiCN nanoparticles transited at an addition amount of 20 wt % and became severe at an addition amount of 40 wt %. For the five Al_2O_3-TiCN composites, the maximum size of TiCN nanoparticle was lower than 150 nm.

In summary, the Al_2O_3-TiCN composites with 5 and 10 wt % TiCN had well-dispersed TiCN nanoparticles and A20T, A30T and A40T composites had agglomerated TiCN nanoparticles.

The microhardness and indentation toughness of A5T and A10T composites were higher than that of monolithic Al_2O_3 ceramic. This indicates the two composites had good mechanical property, see Fig. 3. The microhardness of Al_2O_3-TiCN composites with higher mass fraction of TiCN were lower than 20 GPa and not considered as materials with good mechanical property.

In summary, A5T and A10T composites had good microstructure for good mechanical property. The mechanical property of A20T composite might be improved by optimizing the sintering parameters.

3 Tribological Property and Wear Mechanism of Al_2O_3-TiCN Composite

3.1 Experimental

3.1.1 Tribological Tests

Tribological tests were conducted on a THT high temperature tribometer (CSM Instrument Ltd., Switzerland) with a pin-on-disk configuration at room temperature and 500 °C in air. Friction coefficient was automatically recorded by the computer. The test condition is 5 N for normal load and 0.5 m/s for sliding speed and 1 km for sliding distance. The contact between the hemispherical tip of the pin and the flat surface of the disk is believed to be stable and enables good repeatability of the test.

The pin was made of a Ni-20 wt % Cr alloy (Ni-Cr alloy) prepared by hot pressing at 1,200 °C and 15 MPa for 15 min. The Ni-Cr alloy pin had a size of 6 mm in diameter and 12 mm in length. One end of the pin was machined into a hemispherical tip with a radius of 6 mm for the sliding contact. The surface roughness Ra of the ground hemispherical tip was lower than 0.3 μm. The disk was made of monolithic Al_2O_3 ceramic or Al_2O_3-TiCN composite with a size of 25 mm in diameter and 8 mm in thickness. The surface roughness Ra of the polished disk was lower than 0.1 μm. Both the pin and disk were ultrasonically cleaned in an ethanol bath and allowed to dry prior to the tribological test.

The wear volume of the pin was determined by measuring the wear scar diameter of the pin. The 3D topography of the worn surface of monolithic Al_2O_3 ceramic and Al_2O_3-TiCN composite was observed on a Micro XAM Interferometric Surface Profile (ADE, USA). The wear volume and surface profile of the cross-sectioned wear scar of the disk can be obtained accordingly.

3.1.2 Characterization of Worn Surface

The morphology of the worn surface of monolithic Al_2O_3 ceramic and Al_2O_3-TiCN composite was observed on a SEM (JSM 5600LV, Japan) using secondary electron image (SEI) and backscattered electron image (BEI). The chemical state of typical elements on the worn surface of monolithic Al_2O_3 ceramic and Al_2O_3-TiCN composite was determined by X-ray Photoelectron Spectroscopy (XPS, PHI-5702, USA).

3.2 Results

3.2.1 Tribological Behavior

Room temperature. Friction coefficients of monolithic Al_2O_3 ceramic and Al_2O_3-TiCN composites in sliding against Ni–Cr alloy at room temperature were as high as around 0.7, see Fig. 4a. However, the wear rates of monolithic Al_2O_3 ceramic,

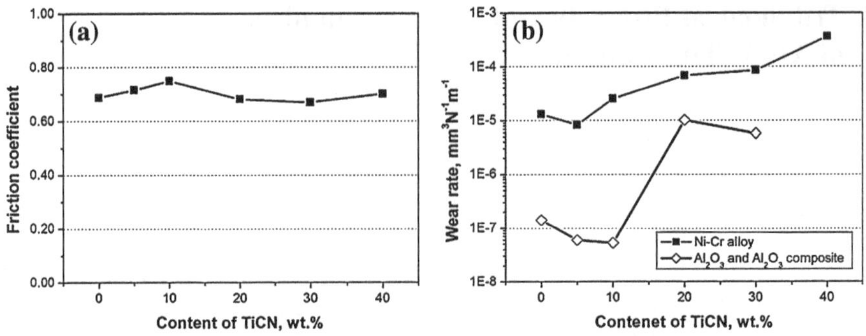

Fig. 4 **a** Friction coefficient and **b** wear rates of Al_2O_3-Ti(CN) composite in sliding against Ni-Cr alloy at room temperature in air. The wear rate of A40T composite is extremely high and not plotted in the figure

Al_2O_3-TiCN composites and their counterpart Ni-Cr alloy ranged from 10^{-8} to 10^{-4} mm^3/(N.m), see Fig. 4b. In addition, the wear rates of monolithic Al_2O_3 ceramic and Al_2O_3-TiCN composites were much lower than that of corresponding Ni-Cr alloy. For example, the wear rates of monolithic Al_2O_3 ceramic and Ni-Cr alloy were on the order of magnitude of 10^{-7} mm^3/(N.m) and 10^{-5} mm^3/(N.m), respectively. Compared with monolithic Al_2O_3 ceramic, the wear rates of Al_2O_3-5 wt % TiCN composite and its counterpart material were lower. The wear rate of A10T composite was even lower but the wear rate of Ni-Cr alloy was higher. The wear resistances of Al_2O_3-TiCN composites (20 wt % and higher content of TiCN) and Ni-Cr alloy were very poor, especially for the composite.

High temperature at 500 °C. Friction coefficients of monolithic Al_2O_3 ceramic and Al_2O_3-TiCN composites in sliding against Ni-Cr alloy at 500 °C were lower than that at room temperature, i.e. 0.2 for A10T and A20T composite; 0.4 for the rest materials, see Fig. 5a. Therefore, the A10T and A20T composites were high temperature self-lubricating materials. The wear rates of Ni-Cr alloys in sliding against these two composites were on the order of magnitude of 10^{-7} to 10^{-6} mm^3/(N.m) and lower than that against the rest materials, see Fig. 5b. Negative wear, which was the result of material transfer from Ni-Cr alloy, was found for monolithic Al_2O_3 ceramic and Al_2O_3-TiCN composites, see Fig. 5b. In addition, the absolute value of the negative wear is an indication of the amount of transferred material. It is interesting that the volume of the transferred material on A20T composite was about 10 times as high as that on A10T composite despite the two composites had identical friction coefficient.

The above results in Figs. 4 and 5 indicate that the tribological behaviors and wear mechanisms of monolithic Al_2O_3 ceramic and Al_2O_3-TiCN composites in sliding against Ni-Cr alloy depend on testing temperature and content of TiCN. As a good example, the frictional traces of A10T composite in sliding against Ni-Cr alloy at room temperature and 500 °C indicate different tribological

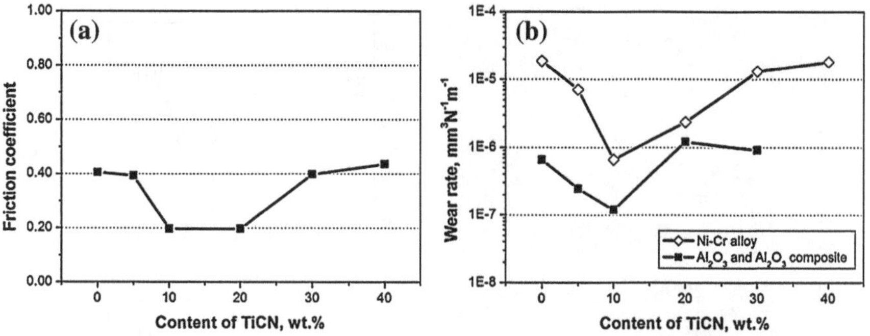

Fig. 5 **a** Friction coefficient and **b** wear rate of monolithic Al_2O_3 ceramic and Al_2O_3-Ti(CN) composite in sliding against Ni-Cr alloy at 500 °C in air. The wear rates of monolithic Al_2O_3 ceramic and Al_2O_3-TiCN composites were negative but presented in absolute value just for better reading

Fig. 6 Typical frictional traces of A10T composite in sliding against Ni-Cr alloy at room temperature and 500 °C

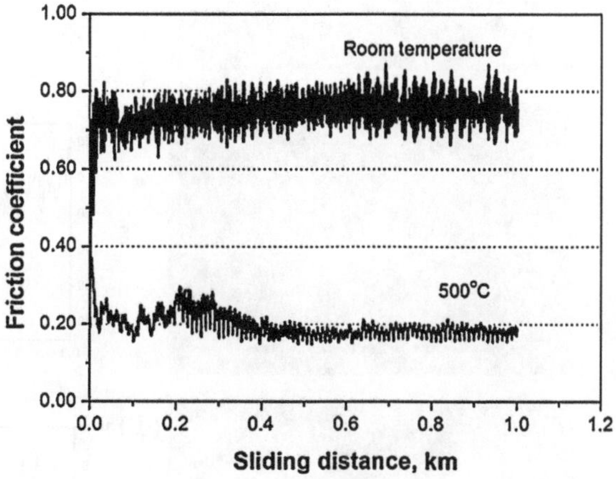

behavior, see Fig. 6. In the following section, the wear mechanisms at room temperature and 500 °C will be separately discussed.

3.2.2 Abrasive Wear and Transfer at Room Temperature

Abrasive wear and transfer (adhesive wear) were the two main wear mechanisms of monolithic Al_2O_3 ceramic and Al_2O_3-TiCN composites in sliding against Ni-Cr alloy at room temperature.

Abrasive wear can be clearly found by observing the 3D morphology and line profile of the wear tracks of monolithic Al_2O_3 ceramic and Al_2O_3-TiCN composites, see Fig. 7. The abrasive particles was generated from monolithic Al_2O_3 ceramic and Al_2O_3-TiCN composites. The resistances to abrasion of monolithic Al_2O_3 ceramic and Al_2O_3-TiCN composites (5 and 10 wt % TiCN) were excellent as evident by shallow scratch marks in Fig. 7a, a', b, b', c, and c'.

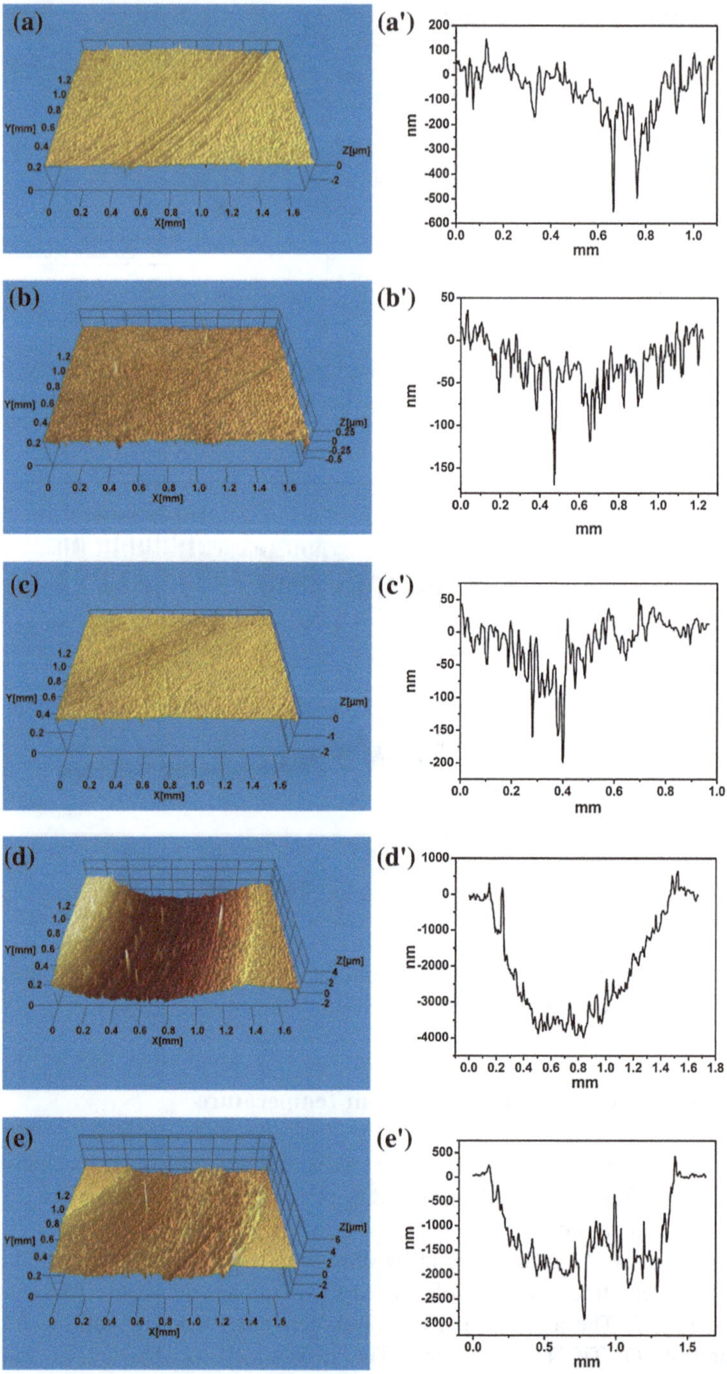

Fig. 7 3D morphologies of the wear tracks of **a** monolithic Al_2O_3 ceramic, **b** A5T composite, **c** A10T composite, **d** A20T composite, **e** A30T composite at room temperature and corresponding line profiles of **a'** monolithic Al_2O_3 ceramic, **b'** A5T composite, **c'** A10T composite, **d'** A20T composite, **e'** A30T composite

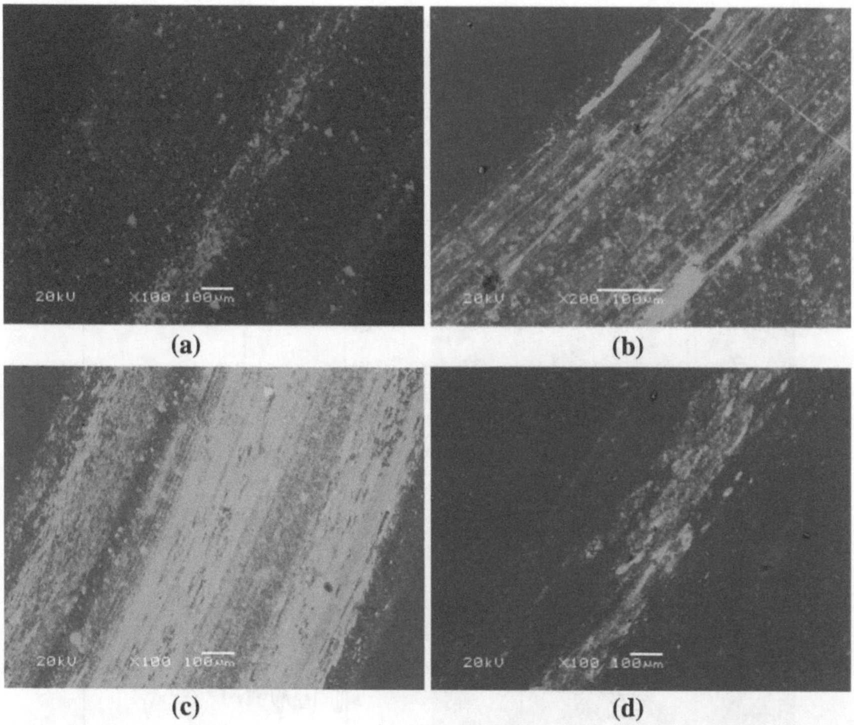

Fig. 8 BEI images of the worn surfaces of **a** monolithic Al_2O_3 ceramic and **b** A5T composite at room temperature and **c** monolithic Al_2O_3 ceramic and A10T composite at 500 °C. The transfer layer is located at the white area

The depth of the scratch mark on monolithic Al_2O_3 ceramic was 300 nm. Due to the reinforcement from hard TiCN nanoparticles in Al_2O_3 matrix, the depth of the scratch marks on A5T and A10T composites were 150 and 100 nm. The agglomerated TiCN nanoparticles in A20T composite (Fig. 2d) were easily pull out as abrasive particles and the resistance to abrasion of the composite was very poor, see Fig. 7d and d'. This is the same for A30T composite, see Fig. 7e and e'.

The transfer layers on monolithic Al_2O_3 ceramic and Al_2O_3-TiCN composites can be identified by BEI image (Fig. 8a). The transfer can be enhanced by the TiCN nanoparticles in Al_2O_3 matrix, see Fig. 8b.

3.2.3 Transfer and Tribo-Oxidation at 500 °C

For monolithic Al_2O_3 ceramic, transfer was greatly enhanced at 500 °C and obviously severer than that at room temperature, see Fig. 8c. The highest asperity on the worn surface of monolithic Al_2O_3 ceramic at 500 °C can be as high as 1.2 μm,

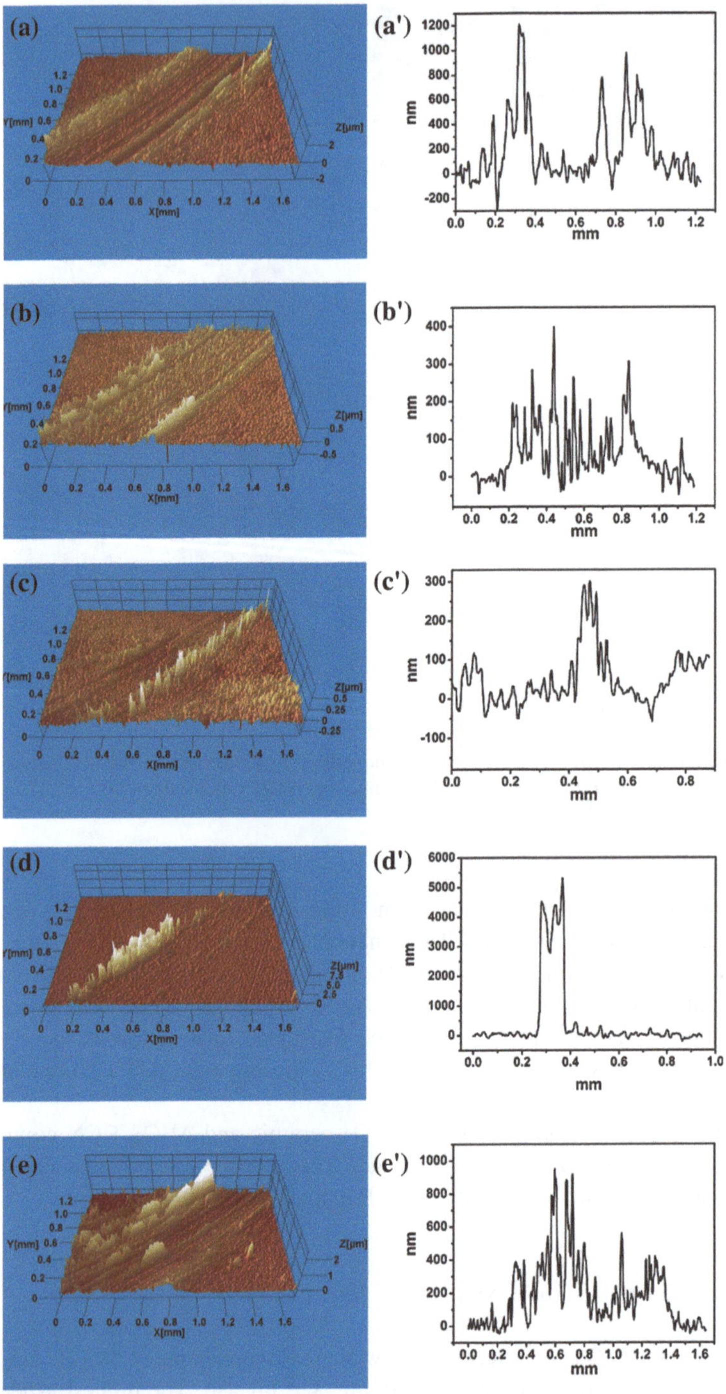

Fig. 9 3D morphologies of the wear tracks of **a** monolithic Al_2O_3 ceramic, **b** A5T composite, **c** A10T composite, **d** A20T composite, **e** A30T composite at 500 °C and corresponding line profiles of **a'** monolithic Al_2O_3 ceramic, **b'** A5T composite, **c'** A10T composite, **d'** A20T composite, **e'** A30T composite

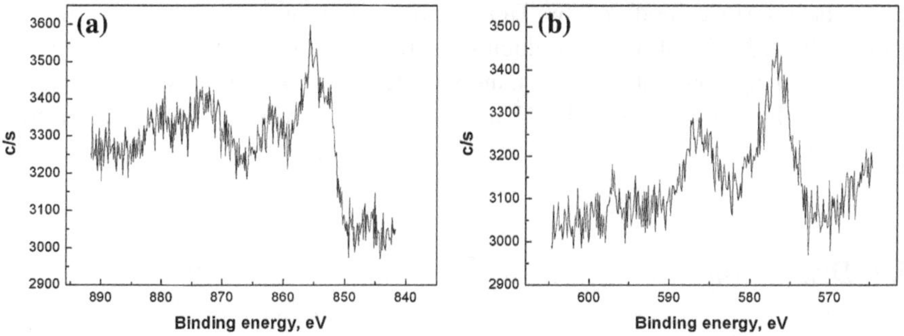

Fig. 10 XPS spectra of **a** Ni2p, and **b** Cr2p on the worn surface of monolithic Al_2O_3 ceramic at 500 °C

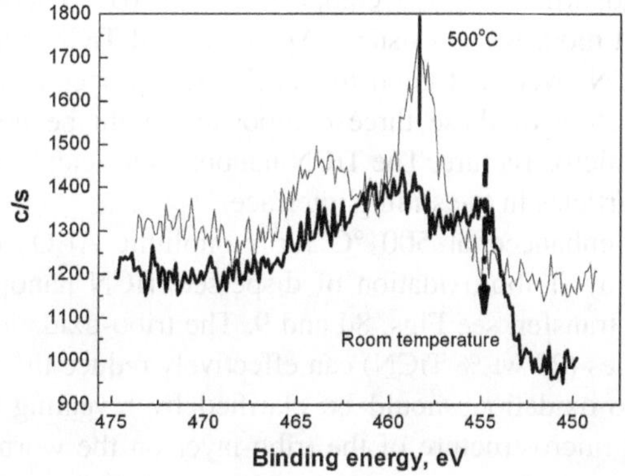

Fig. 11 XPS spectra of Ti2p on the worn surface of A10T composite at room temperature and 500 °C in air

see Fig. 9a and a'. XPS spectra in Fig. 10 reveal that the transfer layer on monolithic Al_2O_3 ceramic at 500 °C was composed of Ni, NiO and Cr_2O_3. As such, the tribological contact was not a pure metal to metal contact, and NiO was responsible for the lower friction coefficient (0.4) than that (0.7) at room temperature.

According to Fig. 5b, the amount of transferred material on Al_2O_3-TiCN composites depended on the content of TiCN nanoparticles in the composite. Fig. 9 clearly demonstrates the transfer layer on the worn surfaces. It is interesting that A5T and A10T composites had small amount of transferred material but quite different friction coefficient. Meanwhile, A10T composite and A20T composite had quite different amount of transferred material but identical friction coefficient.

Tribo-oxidation of TiCN nanoparticles at 500 °C was the key to understand the above results. TiCN nanoparticles can be statically oxidized at 500 °C in air. The tribo-oxidation readily occurred at the sliding interface, see Fig. 11 and the tribo-product TiO_2, which is a softer oxide in comparison with Al_2O_3, modified the chemical composition and microstructure of the tribo-interface.

The tribo-interface modified by TiO_2 played important roles in preventing the transfer from Ni-Cr alloy and reducing friction coefficient at 500 °C. The tribo-oxidation of agglomerated TiCN nanoparticles in A20T composite took advantage over that of dispersed TiCN nanoparticles in A5T composite in friction reduction.

3.3 Discussion

In Sect. 3.2, the tribological behaviors of Al_2O_3-Ti(CN) composites with dispersed and agglomerated TiCN nanoparticles in sliding against Ni-Cr alloy at room temperature and 500 °C in air were investigated. At room temperature, dispersed TiCN nanoparticles in A5T composite and A10T composite made the composite harder and more wear resistant. Agglomerated TiCN nanoparticles (20, 30 and 40 wt % TiCN) were not good for mechanical property and wear resistance. Sintering temperature of these three composites might be not high enough for making a dense microstructure. The TiCN nanoparticles can be readily pullout and act as abrasive particles in the sliding interface.

Transfer was enhanced at 500 °C for monolithic Al_2O_3 ceramic in sliding against Ni-Cr alloy. Tribo-oxidation of dispersed TiCN nanoparticles can effectively inhibit the transfer, see Figs. 8d and 9. The tribo-oxidation of agglomerated TiCN nanoparticles (20 wt % TiCN) can effectively reduce the friction coefficient. The role of tribo-oxidation should be clarified by revealing both the chemical composition and microstructure of the tribo-layer on the worn surface of Al_2O_3-Ti(CN) composites.

Lubricious oxide formed by tribo-oxidation of the second phase nanoparticles at high temperature proves to be effective for A10T and A20T composites. Lubricious double oxides can be generated on the fractional surface of Al_2O_3-based nanocomposites by using a third phase. Lubricious double oxides may provide good lubrication at a higher temperature or wider temperature range [9].

3.4 Conclusions

A concept for designing and fabricating high temperature self-lubricating Al_2O_3-based nanocomposites is proposed using Al_2O_3-Ti(CN) composites as an example. Al_2O_3-Ti(CN) composites with dispersed and agglomerated TiCN nanoparticles were successfully prepared by hot pressing at 1,400 °C. Al_2O_3-Ti(CN) composites with dispersed TiCN nanoparticles have high hardness and good resistance to abrasive wear at room temperature. Al_2O_3-Ti(CN) composites with dispersed (10 wt %) and agglomerated (20 wt %) TiCN nanoarticles are self-lubricating in sliding against Ni-Cr alloy at 500 °C.

Acknowledgments The authors acknowledge for the financial support from National Science Foundation of China (51075382) and One-hundred Talent Project (Junhu Meng) of Chinese Academy of Sciences.

References

1. Jahanmir S (1994) Friction and wear of ceramics. Marcel Dekker Inc, New York
2. Niihara K (1991) New design concept of structural ceramics-ceramic nanocomposite. J Ceram Soc Jpn 99:974–982
3. Sekino T et al (1997) Reduction and sintering of a nickel-dispersed-alumina composite and its properties. J Am Ceram Soc 80:1139–1148
4. Limpichaipanit A, Todd RI (2009) The relationship between microstructure, fracture and abrasive wear in Al_2O_3/SiC nanocomposites and microcomposites containing 5 and 10 % SiC. J Eur Ceram Soc 29:2841–2848
5. Ortiz-Merino JL, Todd RI (2005) Relationship between wear rate, surface pullout and microstructure during abrasive wear of alumina and alumina/SiC nanocomposites. Acta Mater 53:3345–3357
6. Shapiro IP et al (2011) An indentation model for erosive wear in Al_2O_3/SiC nanocomposites. J Eur Ceram Soc 31:85–95
7. Chen HJ et al (2000) The wear behaviour of Al_2O_3-SiC ceramic nanocomposites. Scripta Mater 42:555–560
8. Lu JJ et al. (2010) Friction and wear of Al_2O_3-Ni composite. In: Davim JP (ed) Tribology of composite materials. Nova Science Publishers Inc, New York
9. Erdemir A (2000) A crystal-chemical approach to lubrication by solid oxides. Tribol Lett 8:97–102

Wear of Multi-Scale Phase Reinforced Composites

Zhenyu Jiang and Zhong Zhang

Abstract Multi-scale phase reinforced composites (MPRCs) have been undergoing accelerating progress in the past decade. A judicious combination of micron-scale fibers and nano-scale fillers, such as nanoparticles and nanotubes, are found to endow polymer composites enhanced mechanical and wear performance superior to conventional fiber-reinforced composites or nanocomposites. This chapter reviews the research on the wear and tribological behaviours of the MPRCs reinforced with short fibers and nano-fillers. The dramatic improvements reported in wear properties of MPRCs are considered principally ascribable to a new wear mechanism associated with the interaction between short fibers and nanoparticles. The rolling effects of nanoparticles which protect the fibers and reduce the severity of the wear of MPRCs are elucidated based on experimental findings. By using artificial neural network technique, a phenomenological model study can be carried out on the wear behaviours of MPRCs, which are complicated and influenced by manifold factors.

1 Introduction

The attraction of polymer composites lies in the combination of advantages of various materials, i.e. polymeric matrix and reinforcing phases. Those merits of polymer composites, such as flexibility for design and manufacturing, competitive

Z. Jiang (✉)
Department of Engineering Mechanics, South China University of Technology,
381 Wushan Road, 510641 Guangzhou, China
e-mail: zhenyujiang@scut.edu.cn

Z. Zhang
National Center for Nanoscience and Technology, No.11 ZhongGuanCun BeiYiTiao,
100190 Beijing, China

J. Paulo Davim (ed.), *Tribology of Nanocomposites*, Materials Forming,
Machining and Tribology, DOI: 10.1007/978-3-642-33882-3_5,

cost, excellent performance and chemical resistance, are hardly attained by metals, ceramics, or polymers alone. Among the polymer composites, fiber reinforced polymer composites (FRPs) have been developed for several decades and been used in numerous applications due to their outstanding mechanical and tribological performance. Since the 1990 s, tremendous progress has been achieved in nano-scale filler (e.g. nanoclay, nanotubes and nanoparticles) reinforced polymer composites, which is considered as promising structural and multifunctional engineering materials. Therefore, it is a natural step to combine the two kinds of exceptional reinforcements and build a multi-scale phase reinforced composite (MPRC) system.

The MPRCs are also called hybrid or ternary nanocomposites, or fiber-reinforced nanocomposites in literature [1, 2]. Compared with their two-phase counterparts consisting of single-scale reinforcements (fibers or nanofillers) and matrix, MPRCs are three-phase composites tailored with micron-scale fibers, nano-scale fillers and polymer matrix. The research and development of MPRCs was initiated from the midst of 1990s, when researchers attempted to treat the surfaces of reinforcing fibers with carbon nano-fibers [3] and to modify polymer matrix with inorganic nanoparticles [4] in order to obtain further improvements in the mechanical properties of FRPs. The growth of research interest on MPRCs started to accelerate almost a decade later, i.e. since middle 2000s. Up to today, considerable amount of research effort has been directed to the processing and property study of MPRCs, with focus on their mechanical and wear performance. The reported work has proven this type of composites could be a superior competitor to current FRPs and polymer nanocomposites.

Nowadays, polymer composites are frequently used as tribo-engineering materials, replacing metallic materials in growing technical applications where wear and friction are regarded as critical issues. Short fibers are found to be efficient load bearing components [5], which increase the compressive strength and creep resistance of the polymer matrix, leading to enhanced wear performance. Contribution of other fillers, especially in the form of particulate, to the tribological properties of polymer composites is more complicated. Some "soft" fillers, such as Polytetrafluoroethylene (PTFE) particles and graphite flakes, are extensively used as internal lubricants to alleviate the adhesion between the sliding bodies. The corresponding reduction in the frictional coefficient promoted by these lubricants is attributed to the formation of smooth films on the surfaces of composite and counterpart. As a result, wear and frictional heating are markedly decreased [6, 7]. Inorganic micron-scale particles, harder and stronger than polymer matrix, show various effects, including both positive and negative, on the wear performance of polymer composites. For example, the wear rate of polyphenylene sulphide (PPS), a major high performance thermoplastic, was reported to be reduced by Ag_2S, CuS, NiS, SiC and Cr_3C_2, but increased by PbTe, PbSe, ZnF_2, SnS and Al_2O_3 particles [8–13]. It was found that some beneficial particles are of help to ameliorate the bonding between the transfer film and the metallic counterpart, resulting in improved wear resistance [14], whereas the deleterious particles usually cause thick and lumpy transfer films which have the tendency to detach itself from the counterface.

Nanoparticles with large specific surface area demonstrate dramatic efficiency at reinforcing the mechanical and tribological properties of polymer matrix in comparison with their micrometer counterparts. Furthermore, the action of particle angularity, which is detrimental to the wear performance of composites, can be substantially diminished on nano-scale. However, both positive and negative influences on the tribological properties of PPS have also been observed for different nano-particles. Some contribute towards the formation of thin and uniform transfer film with better adhesion to the counterface, whereas others act as abrasive body only and lead to more severe wear [15, 16].

What makes MPRCs intriguing is the synergistic effect imposed by nanoparticles and short fibers. The wear resistance of MPRCs containing short fibers and nanoparticles could be significantly improved due to the interaction between these reinforcements of multiple scales during the sliding. Zhang et al. [17] reported in 2004 that the specific wear rate of epoxy-based MPRC was drastically reduced to a level about 100 times lower than that of neat epoxy by a combination of micron-scale short carbon fibers, Graphite flakes and PTFE particles, together with nano-TiO_2 particles. The wear resistance of MPRC was found even pronouncedly better than conventional FRP containing internal lubricants. Since then, wear of MPRCs has become a subject of increasing interest to academic and industrial communities.

This chapter presents a review on the progress of the development and investigation of MPRCs with short fibers as primary reinforcement for tribological applications. The discussion covers the preparation techniques and wear behaviours of MPRCs, as well as the new wear mechanisms introduced by the synergistic effect of short fibers and nanoparticles.

2 Experimental Details

2.1 Preparation of Multi-Scale Phase Reinforced Composites

Figure 1 schematically illustrates the procedure for producing MPRCs, in which fillers of multiple scales are introduced to modify the properties of polymers. According to open literature, short carbon fibers (SCFs) [17–34], short glass fibers (SGFs) [35] and short aramid fibers (SAFs) [36–38] are used as primary reinforcement. The internal lubricants, if added, include micrometer sized PTFE particles or graphite flakes [17, 19–21, 25, 26, 28, 29, 33]. Inorganic nanoparticles are most widely used nano-filler, whereas carbon nanotubes [22, 23] and nanoclay [18, 35] are seldom adopted, probably due to the difficulty in filler dispersion.

It is well known that random and homogeneous dispersion of nano-fillers (carbon nanotubes in particular) still remains a challenge so far, and agglomeration of nano-fillers is considered as one of the formidable obstacles in gaining the desired properties of polymer nanocomposites. Therefore, for thermoset-based MPRCs, the multi-scale components are generally incorporated into the matrix separately, from

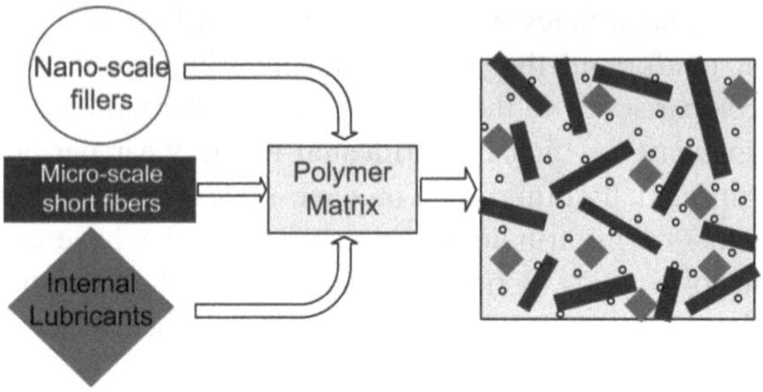

Fig. 1 Schematic diagrams of the procedure for the fabrication of MPRCs. Please note the diagrams are not to scale

nano-scale to micron-scale, in order to achieve a uniform dispersion of all kinds of fillers within the matrix. A typical procedure is summarized as three or four steps, depending on the additional filler besides short fibers and nanoparticles [34]:

1. Nanoparticles are dispersed into the preheated resin of low viscosity through mechanical stirring in a dissolver, occasionally with the aid of ultrasonication to break the agglomerates;
2. If the composition includes internal lubricant, a similar process is performed for the addition of micron-scale lubricant fillers.
3. The nanocomposite resin and short fibers are mechanically mixed after the solvent (if used) is evacuated under vacuum circumstance;
4. Then the hardener agent is added into the mixture, followed by degassing, casting and curing of MPRC specimens.

The preparation of thermoplastic-based MPRCs is restricted to melt blending approaches. In a typical procedure, the nano-fillers, lubricants, short fibers and matrix (in the forms of powder or granule) are mixed together using direct mixing techniques, e.g. mechanical stirring and ultrasonication [36, 39]. Afterward the mixture is melt compounded to MPRCs, through single/twin-screw extrusion, kneading, or compression molding. The procedure could be simplified further for the researchers, because some pre-mixed thermoplastic products containing short fibers and internal lubricants (in the form of granule) are commercially available. In that case, the MPRCs can be prepared by melt blending after simple mixing of the composite granules and nano-fillers [33].

The major advantages of this matrix-modify route for fabrication of MPRCs lie in its simplicity and cost efficiency. The procedure is highly compatible with current industrial techniques and facilities for large-scale manufacturing of polymer composites, which outlines a promising picture for extensive tribological applications of MPRCs in near future.

2.2 Wear Tests and Characterization

Wear resistance, as a most critical performance indicator for tribo-engineering materials, can be assessed by various sliding tests, among which pin-on-disc (POD)

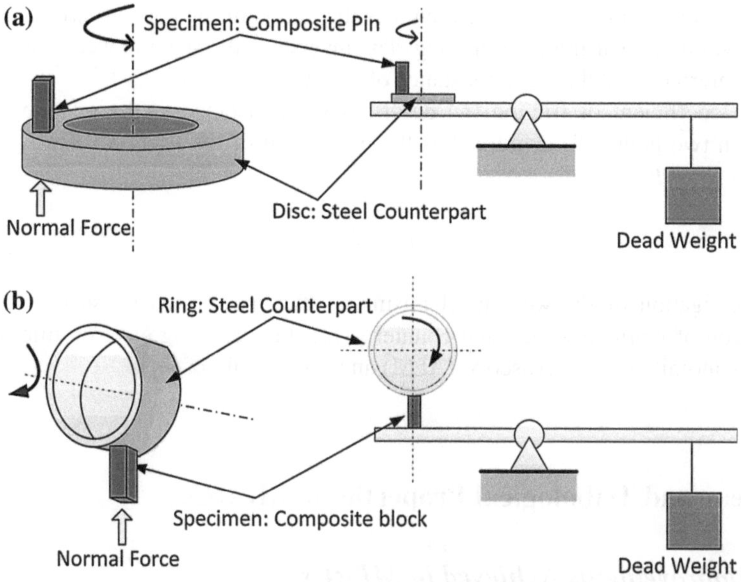

Fig. 2 Illustration of the apparatuses for wear test: **a** pin-on-disc; **b** block-on-ring

and block-on-ring (BOR) are two commonly adopted approaches. Figure 2 shows the schematic diagrams of the apparatuses for the two kinds of tests. The readers can refer to relevant standards, e.g. ASTM G 99 and ASTM G 77, for more details of configuration. In those testing methods, two characteristic parameters are employed to describe the wear and tribological properties of a tribo-material: specific wear rate and coefficient of friction.

Specific wear rate (w_s) represents the volume loss of a substance after wear experiments, normalized by the sliding conditions, i.e. normal force and sliding distance. For pin-on-disk test, it is calculated using the equation:

$$w_s = \frac{\Delta m}{\rho F_N v t} \tag{1}$$

where Δm is the weight loss of specimen after the test, ρ is the material density, F_N is the normal force imposed to the specimen during sliding, v is the sliding speed, and t is the total time that test lasts. In block-on-ring test the metallic counterpart ring leaves a curved worn surface on the composite specimen, the specific wear rate can be calculated more precisely according to the geometrical measurement of the tested specimen using microscope, according to the equation:

$$w_s = \frac{B}{F_N v t}\left[\frac{\pi r^2}{180}\arcsin\left(\frac{b}{2r}\right) - \left(\frac{b}{2r}\right)\sqrt{r^2 - \frac{b^2}{2}}\right] \tag{2}$$

where B is the width of the specimen, r is the radius of the counterpart ring, and b is the width of wear trace on the ring. The inverse of the specific wear rate is generally referred to as the wear resistance of a material.

The coefficient of friction (f_c) describes the resistance to the relative sliding between two bodies. It is defined as the proportion of the friction force F_F to the normal force F_N, namely

$$f_c = \frac{F_F}{F_N} \tag{3}$$

Investigation of the wear mechanisms is often performed by surface characterization of worn specimen and counterpart, using scanning electron microscope (SEM), atomic force microscope (AFM) and optical microscope.

3 Wear and Tribological Properties of MPRCs

3.1 Improvements Achieved in MPRCs

Combinations of various short fibers and nano-fillers have been examined heretofore for the improvement in wear performance of thermoset- and thermoplastic-based MPRCs. According to these attempts, a judicious configured composition, usually consisting of short fibers, internal lubricants and nano-fillers in an appropriate proportion, demonstrated a great potential to enhance the wear resistance of polymers. Table 1 summarizes the results in the reported work. It can be seen that the additional incorporation of nano-fillers into conventional composites filled with short fibers and internal lubricants significantly reduce the specific wear rate and coefficient of friction further. Nanoparticles can result in the pronounced improvements in wear resistance (33.4–87 %) and in coefficient of friction (36.4–73.5 %). Nanoclay leads to the increase of similar magnitude in wear resistance, viz. 44.4 % further improvement. The reinforcing effect of carbon nanotubes on the SCF reinforced PEEK is insignificant. Only 5.3 and 6.5 % further improvements in wear resistance and coefficient of friction respectively are observed. This may explain why less effort has been dedicated to CNT-doped MPRCs for tribological applications so far.

It is noteworthy that short fiber reinforced composites filled with micron-scale internal lubricant of high content or lubricant oil-load microcapsules can also provide the composites very low wear rate and coefficient of friction. However, such compositions may deteriorate the mechanical properties of the composites since the internal lubricants are generally quite soft materials. This is definitely an undesirable side-effect in most practical applications. Addition of inorganic nanoparticles, which is effective to enhance the mechanical performance of polymer composites, is believed to offset such a detrimental influence induced by lubricant fillers [25, 34].

According to Table 1, the optimum nanoparticle content may fall in a range of 1–5 wt %, depending on the size and nature of the particles. Although in most of

Table 1 Improvements in wear performance of MPRC over conventional composites

Matrix	Nano-filler	Short fiber	Test method	Improvement in wear resistance (optimal content of nano-fillers)	Improvement in coefficient of friction (optimal content of nano-fillers)	Ref. and year
Epoxy	Al_2O_3 (2–10 vol %)	Carbon (10 vol %)	BOR	57 % (5 vol %)	No data	[17] 2004
PPS	CuO (1-4 vol %)	Carbon (5 vol %) Aramid (15 vol %)	POD	87 % (2 vol %) 58 % (2 vol %)	No improvement[a]	[36] 2005
PA 66	TiO_2 (5 vol %)	Carbon (15 vol %)	POD	80.7 % (5 vol %)[b] 41.5 % (5 vol %)[c]	37.1 % (5 vol %)[b] 38.7 % (5 vol %)[c]	[29] 2006
PPS	TiO_2 (1-7 vol %)	Carbon (15 vol %)	POD	50 % (5 vol %)	52 (5 vol %)	[31] 2008
Epoxy	SiO_2 (2–10 wt %)	Carbon (10 wt %)	BOR	80 % (4 wt %)[d]	56 % (4 wt %)[d]	[30] 2009
PEEK	SiO_2 (1.5 wt %)	Carbon (10 wt %)	BOR	56 % (1.5 wt %)	74 % (1.5 wt %)	[26] 2009
PTFE	ZnO (1–5 wt %)	Aramid (15 wt %)	POD	58.1 % (5 wt %)	Deterioration	[37] 2009
PEEK	MWCNT (5 wt %)	Carbon (20 wt %)	BOR	5.3 % (5 wt %)	6.5 % (5 wt %)	[22] 2009
Epoxy PA66	TiO_2 (5 wt %)	Carbon (15 vol %)	POD	77 % (5 wt %) 87 % (5 wt %)	48 % (5 wt %) 56 % (5 wt %)	[19] 2010
PEEK	SiO_2 (1-4 wt %)	Carbon (10 wt %)	BOR	33.4 % (2 wt %)[e]	36.4 % (2 wt %)[e]	[33] 2010
PP	MMT clay (1–5 wt %)	Glass (20 vol %)	POD	44.4 % (5 wt %)[f]	No data	[35] 2011
PEEK	ZrO_2 (5-20 wt %)	Carbon (10 wt %)	POD[g]	57.9 % (20 wt %)	Deterioration	[24] 2012

[a] It was attributed to the dominant lubricating effect of short fibers
[b] Test at high *pv* condition: 8 MPa and 1 m/s
[c] Test at high *pv* condition: 2 MPa and 3 m/s
[d] The total weight fraction of nano-fillers and short fibers was fixed at 10 wt %
[e] Test at high *pv* condition: 7 MPa and 1 m/s
[f] Calculated using weight loss
[g] Test in water

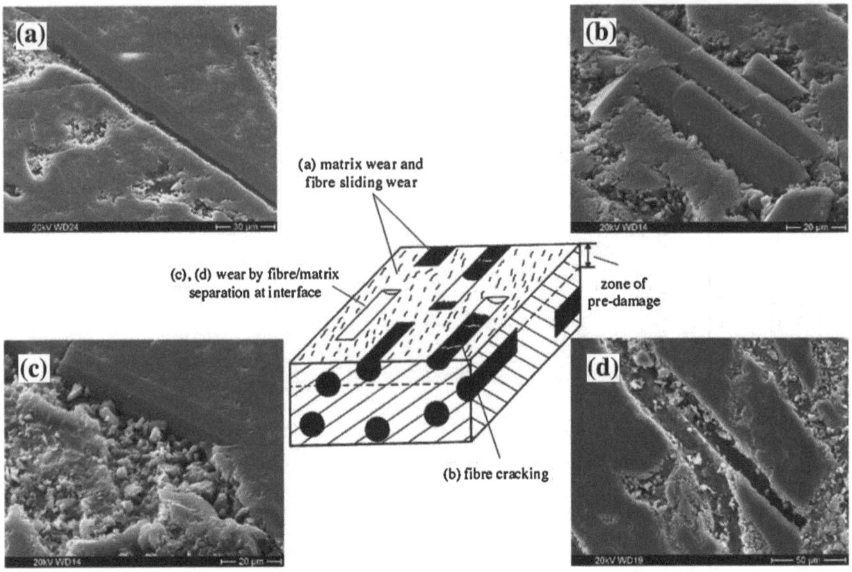

Fig. 3 Schematics of the wear process of short fiber reinforced composite: **a** wear of the matrix and thinning of the fibers; **b** breakage of the fibers; **c** interfacial debonding between fibers and matrix; **d** peeling-off of the fibers [20]

the reported work the particle loadings were seldom over 10 wt % or vol %, and the dispersion quality of nanoparticles were not provided in detail, it is confirmed in the research on the mechanical behaviours of MPRCs that high filler contents tend to result in property deterioration because nanoparticles are prone to form agglomerates under van der Waals force. Those agglomerates, in general act as faults within materials, may affect adversely the wear properties of MPRCs.

3.2 Basic roles of multiple fillers in wear mechanisms

Different types of fillers play their own roles in the conventional composites. Short fibers undertake most of the load due to their high modulus and strength. Figure 3 demonstrates a typical four-stage wear process recognized in short fiber reinforced polymer composites [5, 20]: (1) matrix materials are peeled off during the sliding wear, causing the fibers are exposed to the metallic counterpart; (2) fibers are gradually thinned by the asperities of counterpart until they are broken; (3) debonding between fibers and matrix occurs under concentrated stress around fiber/matrix interfaces; (4) the fibers are pulled out of the matrix, resulting in large cavities.

Internal lubricants lower down the adhesion between the friction bodies. Some layer-structured lubricants, e.g. graphite, contribute to the formation of coherent transfer film on counterpart surface and reduce the coefficient of friction of the paired materials [14], especially in running-in stage of sliding wear test. Accordingly the wear of composites can be substantially diminished.

Micron- and nano-scale particles, when incorporated as sole filler, have manifold influences on the wear behaviours of polymer composites. On the positive side: (1) the hardness, modulus and strength of polymer can be augmented by the addition of hard nano-fillers, thus the wear resistance is enhanced; (2) the wear debris containing particles is helpful to generate smooth and compact transfer film on counterpart surface, which alleviates the friction between the paired bodies. Some particles, such as CuS and CuO [6], are capable to induce tribo-chemical reactions with steel counterpart, whereby the adhesion of transfer film to the counterpart surface is strengthened. On the negative side: (3) Particles, especially in the forms of aggregated clusters, can play a role of abrasive body and aggravate the severity of sliding wear; (4) Moreover, particles may increase the discontinuities in the matrix after they are detached. These adverse effects are inevitable due to the nature of particles and polymers. During the sliding wear of particle-filled composites, the aforementioned beneficial and detrimental effects occur simultaneously and compete against each other. The predominant one determines whether the final wear performance of the composite is upgraded or degraded.

3.3 Rolling Effect of Nanoparticles

What makes MPRCs superior in wear performance to conventional composites (which may also containing multiple fillers) is the combination of multi-scale phases does not mean a simple superimposition of the basic role of each component, but does promote a dramatic synergy between these components, leading to pronounced changes of wear mechanisms.

Fibers as a primary reinforcement in MPRCs carry the load imposed from counterpart, directly via counterpart asperities and indirectly via wear debris. Thus the nanoparticles of much smaller dimension get sufficient support from composite surface. The abrasion of matrix caused by the nanoparticles can be effectively suppressed. Instead, the nanoparticles start to roll rather than slide between the two surfaces. The rolling nanoparticles contribute positively to decreasing coefficient of friction during the sliding, and to lowering down the shear stress level and contact temperature on the surface of MPRCs. In consequence, the wear resistance of MPRCs can be significantly enhanced.

This nano-scale "rolling effect" proposed by Zhang's group [17] has been verified extensively in recent years. One of the strong evidences is the AFM characterization of the worn surface of MPRCs [29] (Fig. 4). Figure 4a shows the fiber in the worn surface of the conventional PA66/SCF composite without nanoparticles. Although the exposed fiber surface is smooth and no fiber/matrix debonding visible in the image, the cross-sectional measurement reveals that the exposed fiber surface is markedly tilted to the worn surface, as shown in Fig. 4c. It suggests that the asperities of metallic counterpart exert an impact load to the front edge of

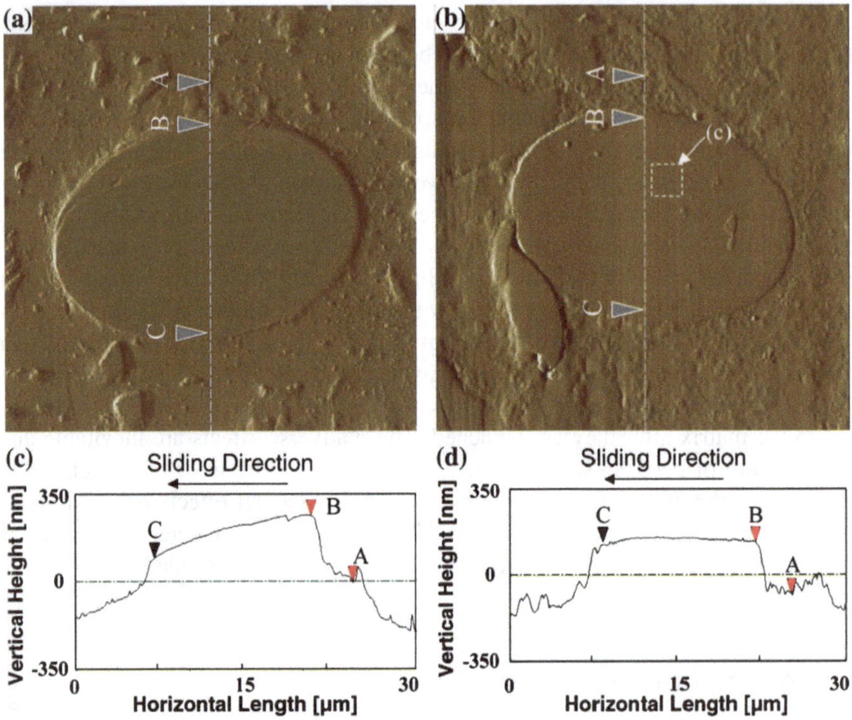

Fig. 4 AFM images of the fiber on the worn surface of the composite **a** without, and **b** with TiO_2 nanoparticles, as well as the corresponding cross-sectional measurements: **c** without, and **d** with nanoparticles. The arrow denotes the sliding direction of counterpart [29]

fiber followed by concentrated press to the surface of fiber, resulting in high shear stress along the fiber/matrix interface, in particular around the front edge against the sliding direction. In remarkable contrast, the fiber surface is well parallel to the worn surface in the PA66/SCF composite with nano-TiO_2 particles (see Fig. 4b and d). The result is consistent with the "rolling effect" in which the nanoparticles act as load transfer body and help to distribute the stress evenly on the fiber surface. By this way, the direct scratch of counterpart asperities is replaced with a relatively milder polish of particles, as illustrated in Fig. 5. Figure 6 shows the SEM photo taken on the worn surface of PA66/SCF/nano-TiO_2 MPRC, where a great amount of shallow nano-scale grooves caused by this nanoparticle polish process is clearly visible on the fiber surface.

Another key role of the rolling nanoparticles is protecting the fibers from being peeled-off from the matrix. As mentioned above, it is the front edge of fibers in conventional short fiber reinforced composites, which takes the brunt of the counterpart asperities. The strong impact upon the polymer matrix in the interfacial area tends to induce fiber/matrix debonding and subsequent removal of the fibers, as illustrated in Fig. 7. In MPRCs, the nanoparticles ploughed out, either operating alone

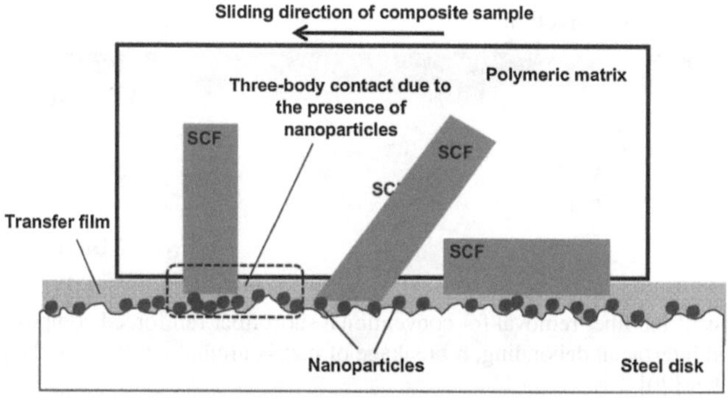

Fig. 5 Illustration of the contact mode for the MPRCs during the sliding [19]

Fig. 6 SEM micrograph of the fiber surface with nano-grooves on the worn surface of PA66/SCF/Nano-TiO_2 MPRC [19]

or being wrapped by the matrix, can be driven by the counterpart asperities and roll between the paired surfaces until they meet the exposed fibers. Then the nanoparticles and wear debris are blocked by the fibers and heaped up, building a protective shield for the front edge of fibers against the direct rubbing imposed by the counterpart, as illustrated in Fig. 8. The wear mechanism is evidenced by SEM photo of the worn surface (Fig. 9). On the worn surface of PPS/SCF/nano-TiO_2 MPRC specimen the accumulation of debris around the front edge of fibers can be clearly observed (Fig. 9b), whereas the front edge of fibers in conventional PPS/SCF keeps very clean (Fig. 9a). This mechanism is critical to enhance the wear resistance of MPRCs, because the fibers can be held in the matrix for longer time until they are cut into very thin pieces finally, and provide a lasting load bearing during the sliding.

A few groups reported that the positive effect of nanoparticles on wear resistance of MPRCs could be amplified under high *pv* conditions (*pv* value means the product of the normal pressure *p* and the sliding velocity *v*) [21, 31, 33, 40].

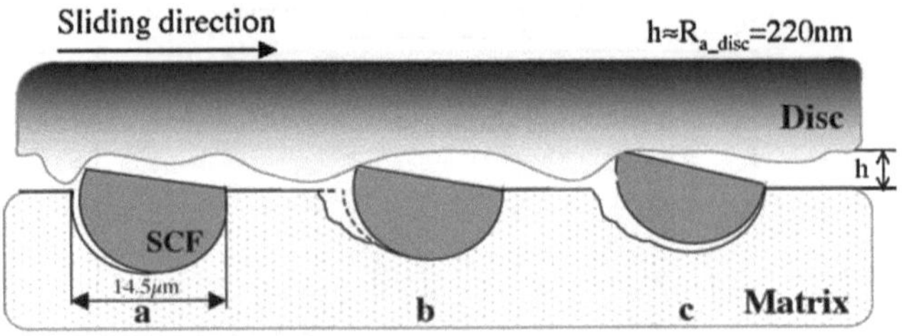

Fig. 7 Process of the fiber removal for conventional short fiber reinforced composites: **a** rotation of the fiber and interfacial debonding, **b** breakage of matrix around interface, and **c** peeling-off of the exposed fiber [20]

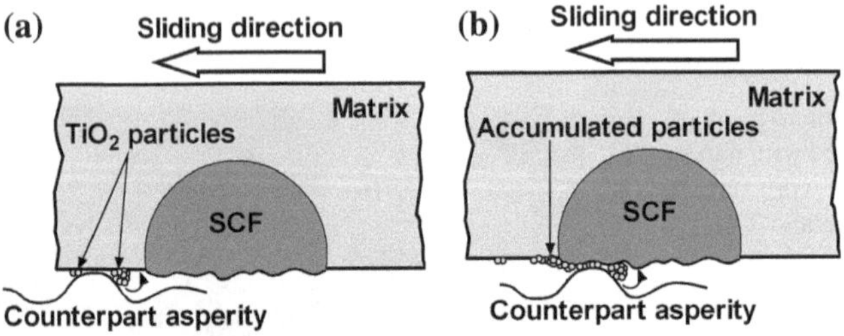

Fig. 8 Diagrams of the protective effect of rolling nanoparticles on the fiber at its front edge: **a** the nanoparticles ploughed out roll with the counterpart asperities; **b** the nanoparticles are blocked by the exposed fiber and accumulate at the front edge of fiber, preventing the counterpart asperity from pulling the fiber out of the matrix. Please not the schematics are not to scale [31]

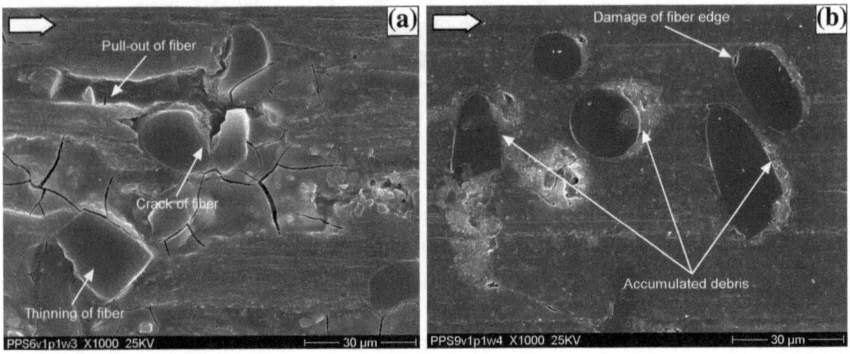

Fig. 9 SEM micrographs of worn surfaces of **a** conventional short fiber reinforced composite and **b** PPS/SCF/nano-TiO_2 MPRC. The arrows in the top-left corner indicate the sliding direction of the specimen [31]

This interesting finding can also be explained by the “rolling effect”. At a high pressure or sliding velocity, the nanoparticle agglomerates of micrometer dimension, acting as abrasive body, are easier to be crushed into smaller scale and shift their role to the rolling third body [33]. More important, the rolling nanoparticles acting as spacers between the composite and counterpart not only reduce the friction, but also suppress the accumulation of heat. As we know most energy dissipated during sliding wear is transformed to heat. A high contact temperature, in particular above the glass transfer temperature of host polymer, decreases the strength of the matrix and induces thermal–mechanical failure, deteriorating the wear resistance of the composite. Under a common testing condition (1 MPa·m/s in air), the contact temperatures of the metallic counterparts for MRPCs are found to be several degrees lower than that for conventional short fiber reinforced composites. The temperature difference can achieve about 20–40 °C degree under testing conditions of high *pv* product (5 MPa·m/s) [28, 29, 33], which make the rolling effect more prominent on lowering down the wear rate of MPRCs. On the other hand, the rolling effect could be weakened if the generated heat is dissipated via other way. For example, the magnitude of improvement in specific wear rate of PEEK/SCF/nano-ZrO_2 MPRCs was observed to decrease from 57.9 % at a condition of 1 MPa and 2 m/s down to 32 % at a condition of 8 MPa and 2 m/s when the sliding wear tests were carried out in water, where the rise of temperature of the water near the paired surfaces was limited less than 10 °C [24].

3.4 Other Reinforcing Mechanisms

Nanoarticles with much higher specific surface area are found to be more effective on reinforcing the mechanical properties of polymer matrix than micron-scale particles. This advantageous feature can be of help to enhance both composite and transfer film in wear.

It has been proven that the addition of nanoparticles results in significant increase in various mechanical properties of polymers, such as modulus, hardness, strength, and fracture toughness. As illustrated in Fig. 10, the nanocomposite matrix with enhanced modulus is considered to give the fibers stronger support when they are rubbed by the counterpart. Diminution of the plastic deformation in zone I (in front of short fiber) and zone II (behind short fiber) conduces towards a tight hold of the fibers within matrix [26].

Considering that the failure of polymer composites during the wear is considerably due to creep and fatigue, the polymer matrix with improved damage tolerance could demonstrate higher wear resistance. Moreover, the stiffening and toughening of polymer matrix by the incorporation of nanoparticles also benefits the interfacial adhesion strength between fibers and matrix [41]. The presence of the nanofillers within the matrix may contribute to a more efficient stress transfer in MPRCs and alleviate the stress concentration along the fiber/matrix interface. Simultaneously, the debonding between the nanoparticles and polymer matrix, and the diverging of the micro-cracks may consume more deformation energy for the

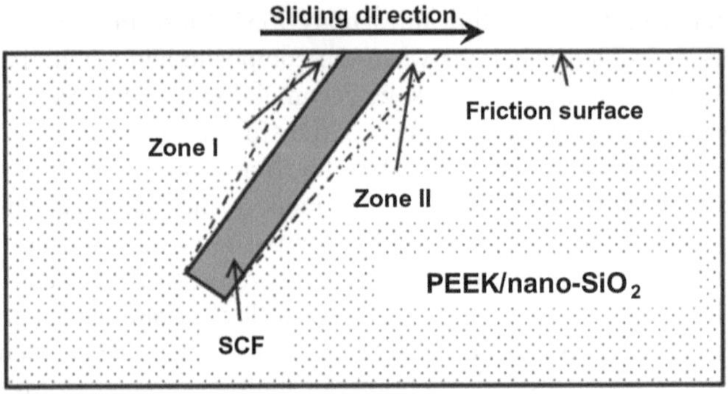

Fig. 10 Illustration of how the enhanced matrix provides a strong support to the short fiber during sliding [26]

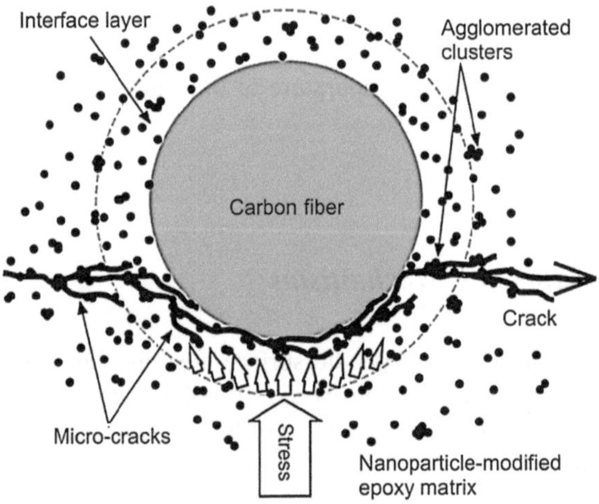

Fig. 11 Illustration of energy dissipation and stress transfer along the fiber/matrix interface in MPRC [41]

development and propagation of micro-cracks, as illustrated in Fig. 11. A strengthened fiber/matrix bonding considerably helps to keep the primary reinforcing fibers in the matrix during the wear.

Transfer film has a pivotal role in protecting the soft polymer surface from the hard metal asperities. According to the study on the transfer film by Bahadur's group [15, 16, 36, 42], properly selected nanoparticles can promote the formation of thin and compact transfer film on the counterpart surface. When the fillers decompose, the product generated in tribo-chemical reaction makes the transfer film more coherent to the counterpart. This coating provides a lasting shielding for the composite against the harsh scraping of the counterpart asperities, reducing the wear rate. Figure 12 compares the transfer films formed in the cases of PPS/SAF

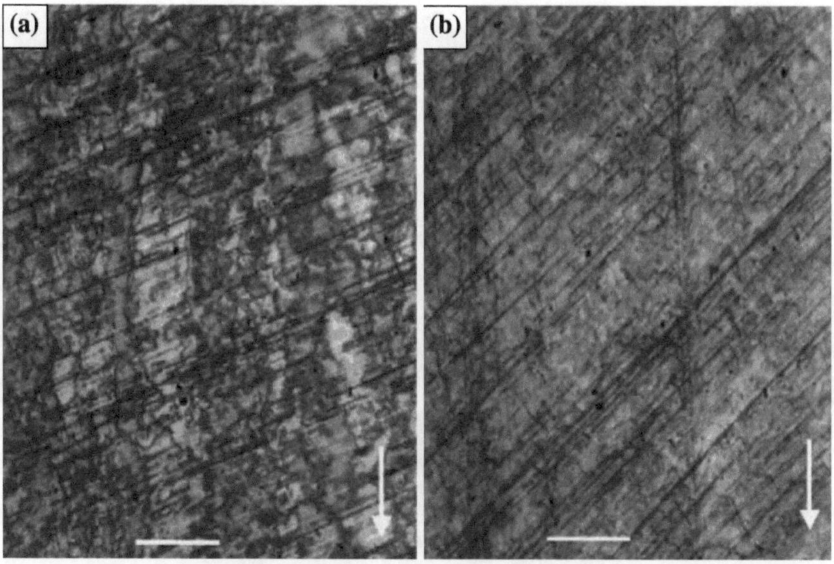

Fig. 12 Optical micrographs of the transfer films on the counterfaces for **a** PPS/SAF composite and **b** PPS/SAF/nano-CuO MPRC. Arrows indicates the sliding direction [36]

composite with and without nano-CuO particles. It can be seen that there are less scratch tracks on the transfer film for PPS/SAF/nano-CuO MPRC than that for PPS/SAF composite, suggesting a milder wear. Figure 13 shows the SEM photos of the transfer films for conventional epoxy/SCF composite, epoxy/SCF MPRCs containing untreated nano-SiO_2 particles and grafted nano-SiO_2 particles (the latter are covalently bonded with matrix after curing) [30]. Apparently the transfer film discontinuously developed during the sliding of epoxy/SCF composites is rather lumpy with relatively deep grooves compared to the cases of MPRCs (Fig. 13a). The addition of nanoparticles contributes positively to the formation of smooth and uniform transfer film (Fig. 13b and c). The epoxy/SCF MPRC filled with grafted nano-SiO_2 particles gives the smoothest transfer film and lowest wear rate accordingly.

4 Modeling of Wear Behaviours for MPRCs

Although the observed amelioration of wear performance in MPRCs can be well explained by the positive rolling effect of nanoparticles, quantitative modeling of the wear properties of MPRCs is still in its infancy stage. Wear of materials is a complex process associated with mechanical interactions, physical transitions and chemical reactions. Moreover, the synergistic effects introduced by the micron-scale short fibers and nanoparticles could be influenced by a great amount

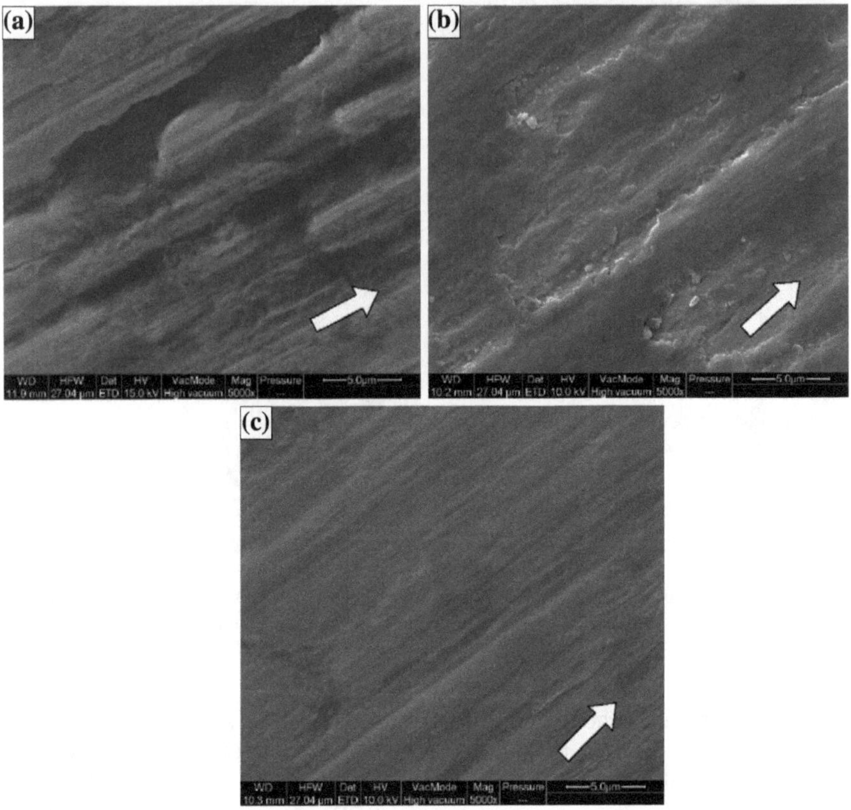

Fig. 13 SEM micrographs of the transfer films on the counterfaces for **a** Epoxy/SCF composite, **b** Epoxy/SCF/untreated nano-SiO_2 MPRC and **c** Epoxy/SCF/grafted nano-SiO_2 MPRC. Arrows indicates the sliding direction [30]

of factors, such as the features of nanoparticls (size, shape, hardness and surface characteristics), dispersion quality of multiple components, and testing conditions. To integrate all of these controlling elements poses a serious challenge to the modeling study.

Artificial neural network (ANN) is a powerful analytical tool for the modeling of complex phenomena. Inspired by biological nerve system, ANN uses inter-connected mathematical nodes to imitate the learning ability of human. Such a network can learn (or be trained) from examples, and explore by itself the underlying functional relationships between the given reasons and results, which generally cannot not be well described by existing physical theories and other mathematical approaches. The knowledge generalized by the network can be applied to make prediction for new cases. Extensive research has proven that ANN is ideally suitable to model the wear, fatigue, and creep behaviours of polymer

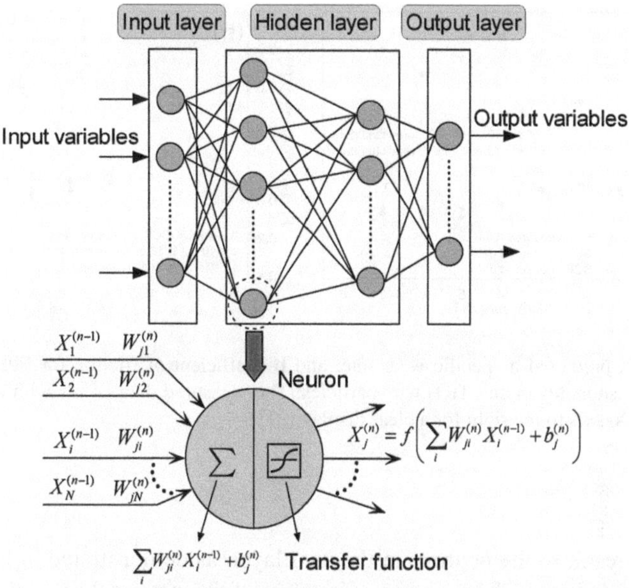

Fig. 14 Structure of artificial neural network [48]

composites [43–47], which represent a type of very complex, non-linear and multi-dimensional problems in materials science.

Figure 14 provides a schematic illustration of an ANN composed of cross-linked mathematic nodes (so-called neurons). An ANN is generally defined as three parts connected in series: input layer, hidden layer and output layer. The raw information is acquired by the input layer, and processed in the hidden layer. Afterward, the results are exported via the output layer. The neuron number in input and output layer are fixed to be equal to the input and output variables respectively, whereas the hidden layer may have more than one layer, and the neuron number in each layer is flexible. The structure of an ANN can be expressed as

$$N_{in} - [N_1 - N_2 - \cdots - N_h]_h - N_{out} \tag{4}$$

where N_{in} and N_{out} are the numbers of input and output variables, respectively. h denotes the number of layers in hidden layer, and N_h is neuron number in the hth hidden layer. For example, 7-[9-3]$_2$-1 represents a network consisting of input layer with 7 input parameters, 2 hidden layers with 9 and 3 neurons respectively, and output layer with 1 output parameter. The neurons, as the basic processing units in a network, are linked by weighted inter-connections that resemble the intensity of the bioelectricity transferring between the neuron cells in a real neural network. During the training process, the individual neuron receives the information from the neurons in the previous layer, and modulates the information with a linear or non-linear transfer function ($f(x)$), weights (W) and bias (b), finally

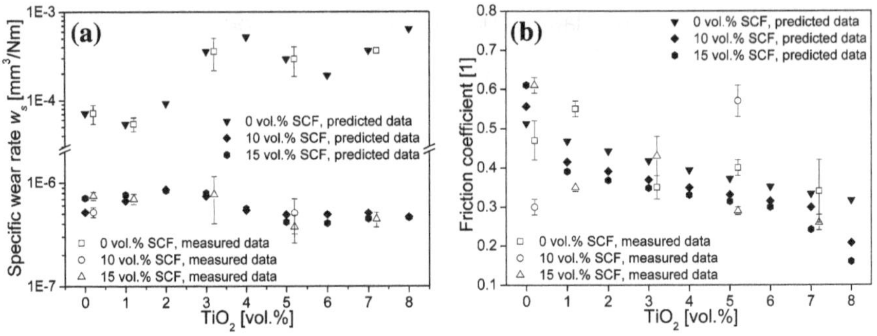

Fig. 15 ANN-predicted **a** specific wear rate, and **b** coefficient of friction for PPS with different content of short fibers and TiO_2 nanoparticles. The measured data points with error bars are shifted along X-axis to the right for a clear display [31]

outputs the result to the neurons in the next layer, as demonstrated in Fig. 14. The learned knowledge can be memorized in terms of the state of these weights as well as the biases in neurons.The application of ANN for the modeling of wear behaviours of polymer composites can be summarized in terms of three stages [48]:

(1) Database collection: collect and pre-process the experimental data to build a large enough database. In this database, any possible controlling fact, e.g. material compositions, manufacturing parameters, testing conditions, and mechanical or physical properties of the composites which may have considerable influences on the wear properties can be used as input, in case the factor can be quantized to certain value. Specific wear rate and coefficient of friction are generally set as outputs.
(2) Optimization of ANN: train the network using a part of the collected database, evaluate its performance using the rest, and then adjust the structure, internal parameters of the network and training algorithm. The procedure is repeated for a few times until the performance of a network, such as prediction accuracy and computing efficiency reach the expected level.
(3) Prediction by ANN: use the sufficiently optimized and trained network to predict solutions for input data which is new to the collected database.

Figure 15 shows the ANN predicted specific wear rate and frictional coefficient of PPS/SCF/nano-TiO_2 MPRCs, accompanied with the measured results (to distinguish the measured data from prediction, those points are shifted a little to right side along X-axis in the figure). It can be seen that ANN prediction gives a good profile for wear rate of MPRCs as function of SCF and nano-TiO_2 contents. Furthermore, an optimum composition of MPRC, namely PPS with 15 vol % SCF and 6 vol % nano-TiO_2 can be estimated according to the ANN model [31]. The predicted frictional coefficients of MPRCs show

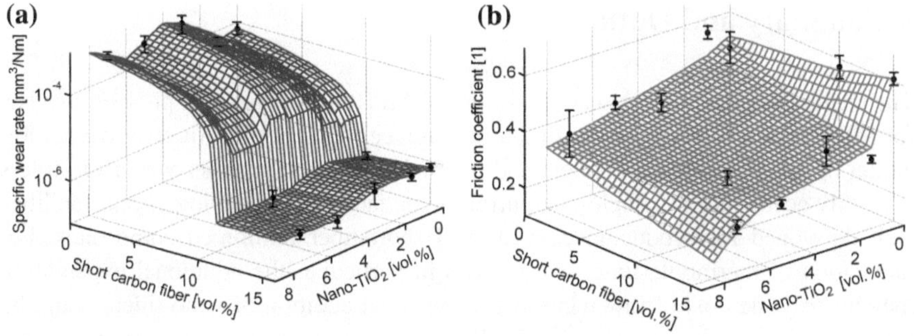

Fig. 16 ANN-predicted 3D profile of **a** specific wear rate, and **b** coefficient of friction as function of the content of multi-scale phases

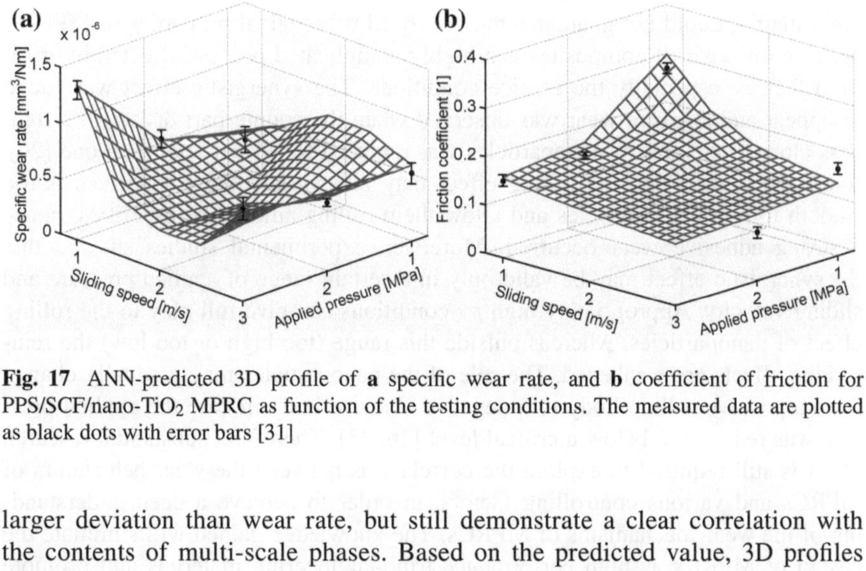

Fig. 17 ANN-predicted 3D profile of **a** specific wear rate, and **b** coefficient of friction for PPS/SCF/nano-TiO_2 MPRC as function of the testing conditions. The measured data are plotted as black dots with error bars [31]

larger deviation than wear rate, but still demonstrate a clear correlation with the contents of multi-scale phases. Based on the predicted value, 3D profiles can be constructed by interpolation to show how the tribological properties of the composites vary in the given ranges of multi-scale phase contents (Fig. 16). The influences of testing conditions, i.e. applied pressure and sliding speed, on wear properties of MPRC can be also well modeled by ANN, as displayed in Fig. 17. The ANN predicted 3D profiles suggest a proper range of service condition for the MPRC.

It should be pointed out that ANN works like a "black box" and unfortunately the explicit model on the physical nature of the wear behaviours cannot be provided by this technique. Although ANN dose not stimulate the understanding of the wear mechanisms in MPRCs, the outstanding ability of this phenomenological modeling approach in exploring the complex relationship between multiple inputs and outputs simultaneously is still of great help to release the researchers of tribo-engineering materials from labor-intensive and time-consuming wear tests.

5 Summary and Outlook

The recent advance in the research on wear of MPRCs has opened a prosperous vista for MPRCs as a type of outstanding tribo-engineering material. An effective combination of micron-scale short fibers and internal lubricants, together with nano-fillers is readily accessible by employing current manufacturing technologies and facilities well developed for producing conventional short fiber reinforced composites. For fabrication of thermoplastic-based MPRCs in particular, the additional expenditure may be only the cost of raw nano-fillers, since the addition of nano-fillers could be merged into the mixing process of other micron-scale components. Furthermore, homogenously dispersed nano-fillers have been extensively reported to make remarkable contribution towards the improvements in mechanical performance of MPRCs, which endow MPRCs with the adaptability to a wider variety of service situations.

On the other hand, multi-scale components, even ideally incorporated into polymer matrix, could not guarantee the improved tribological properties of MPRCs, because the wear of composites is a highly complicated process affected by manifold factors, especially the service conditions. The synergistic effect was found disappear and a severe wear was observed when the counterpart disk with roughness close to the size of nanoparticles was replaced by a finely polished one [20]. It was concluded that the rolling effect may not work if the counterface is too smooth to hold nanoparticles and allow them rolling among the asperities, hence a strong adhesive wear occurred. Moreover, experimental studies showed that the synergistic effect may be valid only in a certain range of applied pressure and sliding velocity. Appropriately high *pv*-conditions can give full play to the rolling effect of nanoparticles, whereas outside this range (too high or too low) the reinforcing effects are weakened. The role of the nanoparticles may gradually change from a positive rolling body to a negative abrasive body in case the applied pressure was reduced to below a critical level [26, 33]. Therefore, substantial research effort is still required to explore the correlations between the wear behaviours of MPRCs and various controlling factors, in order to achieve a deep understanding of the wear mechanisms of MPRCs. The knowledge gained will stimulate the design of MPRCs as high performance tribo-engineering materials and promote their wear-resistant applications.

References

1. Njuguna J, Pielichowski K, Alcock JR (2007) Epoxy-based fibre reinforced nanocomposites. Adv Eng Mater 9(10):835–847
2. Qian H, Greenhalgh ES, Shaffer MSP, Bismarck A (2010) Carbon nanotube-based hierarchical composites: a review. J Mater Chem 20(23):4751
3. Down WB, Baker RTK (1995) Modification of the surface properties of carbon fibers via the catalytic growth of carbon nanofibers. J Mater Res 10(3):625–633
4. Hussain M, Nakahira A, Niihara K (1996) Mechanical property improvement of carbon fiber reinforced epoxy composites by Al_2O_3 filler dispersion. Mater Lett 26(3):185–191

5. Friedrich K (1986) Wear of reinforced polymers by different abrasive counterparts. In: Friedrich K (ed) Friction and wear of polymer composites. Elsevier Science, Amsterdam, pp 233–287
6. Friedrich K, Zhang Z, Schlarb a (2005) Effects of various fillers on the sliding wear of polymer composites. Compos Sci Technol 65(15–16):2329–2343
7. Briscoe BJ, Sinha SK (2008) Tribological applications of polymers and their composites: Past, present and future prospects. In: Friedrich K, Schlarb AK (eds) Tribology of polymeric nanocomposites, vol 55. Elsevier Science, Amsterdam, pp 1–14
8. Cho MH, Bahadur S, Pogosian AK (2005) Friction and wear studies using Taguchi method on polyphenylene sulfide filled with a complex mixture of MoS_2, Al_2O_3, and other compounds. Wear 258(11–12):1825–1835
9. Schwartz CJ, Bahadur S (2001) The role of filler deformability, filler-polymer bonding, and counterface material on the tribological behavior of polyphenylene sulfide (PPS). Wear 251(1–12):1532–1540
10. Yu L, Bahadur S, Xue Q (1998) An investigation of the friction and wear behaviors of ceramic particle filled polyphenylene sulfide composites. Wear 214(1):54–63
11. Yu L, Yang S, Liu W, Xue Q (2000) An investigation of the friction and wear behaviors of polyphenylene sulfide filled with solid lubricants. Polym Eng Sci 40(8):1825–1832
12. Zhao Q, Bahadur S (1998) A study of the modification of the friction and wear behavior of polyphenylene sulfide by particulate Ag_2S and PbTe fillers. Wear 217(1):62–72
13. Zhao Q, Bahadur S (1999) The mechanism of filler action and the criterion of filler selection for reducing wear. Wear 225–229:660–668
14. Bahadur S (2000) The development of transfer layers and their role in polymer tribology. Wear 245(1–2):92–99
15. Bahadur S, Schwartz CJ (2008) The influence of nanoparticle fillers in polymer matrices on the formation and stability of transfer film during wear. In: Friedrich K, Schlarb AK (eds) Tribology of polymeric nanocomposites, vol 55. Elsevier Science, Amsterdam pp 17–34
16. Bahadur S, Sunkara C (2005) Effect of transfer film structure, composition and bonding on the tribological behavior of polyphenylene sulfide filled with nano particles of TiO_2, ZnO, CuO and SiC. Wear 258(9):1411–1421
17. Zhang Z, Breidt C, Chang L, Haupert F, Friedrich K (2004) Enhancement of the wear resistance of epoxy: short carbon fibre, graphite, PTFE and nano-TiO_2. Compos Part A 35(12):1385–1392
18. Basavaraj E, Siddaramaiah (2010) Sliding wear behavior of nanoclay filled polypropylene/ ultra high molecular weight polyethylene/carbon short fiber nanocomposites. J Macromol Sci Part A Pure Appl Chem 47(6):558–563
19. Chang L, Friedrich K (2010) Enhancement effect of nanoparticles on the sliding wear of short fiber-reinforced polymer composites: a critical discussion of wear mechanisms. Tribol Int 43(12):2355–2364
20. Chang L, Zhang Z (2006) Tribological properties of epoxy nanocomposites Part II. A combinative effect of short carbon fibre with nano-TiO_2. Wear 260(7–8):869–878
21. Chang L, Zhang Z, Zhang H, Friedrich K (2005) Effect of nanoparticles on the tribological behaviour of short carbon fibre reinforced poly(etherimide) composites. Tribol Int 38(11–12):966–973
22. Li J, Zhang LQ (2009) Tensile and tribological properties of a short-carbon-fiber-reinforced peek composite doped with carbon nanotubes. Mech Compos Mater 45(5):495–502
23. Li J, Zhang LQ (2011) Reinforcing effect of carbon nanotubes on PEEK composite filled with carbon fibre. Mater Sci Tech-lond 27(1):252–256
24. Lin G-m, Xie G-y, Sui G-x, Yang R (2012) Hybrid effect of nanoparticles with carbon fibers on the mechanical and wear properties of polymer composites. Compos Part B Eng 43 (1):44–49
25. Xian G, Walter R, Haupert F (2006) Friction and wear of epoxy/TiO_2 nanocomposites: Influence of additional short carbon fibers, Aramid and PTFE particles. Compos Sci Technol 66(16):3199–3209

26. Zhang G, Chang L, Schlarb AK (2009) The roles of nano-SiO_2 particles on the tribological behavior of short carbon fiber reinforced PEEK. Compos Sci Technol 69(7–8):1029–1035
27. Zhong YJ, Xie GY, Sui GX, Yang R (2011) Poly (ether ether ketone) composites reinforced by short carbon fibers and zirconium dioxide nanoparticles : Mechanical properties and sliding wear behavior with water lubrication. J Appl Polym Sci 119(3):1711–1720
28. Chang L, Zhang Z, Breidt C, Friedrich K (2005) Tribological properties of epoxy nanocomposites I. Enhancement of the wear resistance by nano-TiO_2 particles. Wear 258(1–4):141–148
29. Chang L, Zhang Z, Zhang H, Schlarb AK (2006) On the sliding wear of nanoparticle filled polyamide 66 composites. Compos Sci Technol 66(16):3188–3198
30. Guo QB, Rong MZ, Jia GL, Lau KT, Zhang MQ (2009) Sliding wear performance of nano-SiO_2/short carbon fiber/epoxy hybrid composites. Wear 266(7–8):658–665
31. Jiang Z, Gyurova L, Schlarb A, Friedrich K, Zhang Z (2008) Study on friction and wear behavior of polyphenylene sulfide composites reinforced by short carbon fibers and sub-micro TiO_2 particles. Compos Sci Technol 68(3–4):734–742
32. Wang Q, Zhang X, Pei X (2010) Study on the synergistic effect of carbon fiber and graphite and nanoparticle on the friction and wear behavior of polyimide composites. Mater Des 31(8):3761–3768
33. Zhang G (2010) Structure—tribological property relationship of nanoparticles and short carbon fibers reinforced PEEK hybrid composites. J Polym Sci Pol Phys 48(7):801–811
34. Guo QB, Lau KT, Rong MZ, Zhang MQ (2010) Optimization of tribological and mechanical properties of epoxy through hybrid filling. Wear 269(1–2):13–20
35. Mohan TP, Kanny K (2011) Influence of nanoclay on rheological and mechanical properties of short glass fiber-reinforced polypropylene composites. J Reinf Plast Comp 30(2):152–160
36. Cho MH, Bahadur S (2005) Study of the tribological synergistic effects in nano CuO-filled and fiber-reinforced polyphenylene sulfide composites. Wear 258(5–6):835–845
37. Wang L-Q, Jia X-M, Cui L, Chen G-C (2009) Effect of aramid fiber and ZnO nanoparticles on friction and wear of PTFE composites in dry and LN_2 conditions. Tribol T 52(1):59–65
38. Tang G, Huang W, Chang D, Nie W, Mi W, Yan W (2011) The Friction and Wear of Aramid Fiber-Reinforced Polyamide 6 Composites Filled with Nano-MoS_2. Polym-plast Technol 50(15):1537–1540
39. Wang Q, Zhang X (2010) A synergistic effect of graphite and nano-CuO on the tribological behavior of polyimide composites. J Macromol Sci Part B Phys 50(2):213–224
40. Wang Q-H, Zhang X-R, Pei X-Q (2010) Study on the friction and wear behavior of basalt fabric composites filled with graphite and nano-SiO_2. Mater Des 31(3):1403–1409
41. Jiang Z, Zhang H, Zhang Z, Murayama H, Okamoto K (2008) Improved bonding between PAN-based carbon fibers and fullerene-modified epoxy matrix. Compos Part A 39(11):1762–1767
42. Schwartz CJ, Bahadur S (2000) Studies on the tribological behavior and transfer film-counterface bond strength for polyphenylene sulfide filled with nanscale alumina particles. Wear 237(2):261–273
43. Zeng P (1998) Neural computing in mechanics. Appl Mech Rev 51(2):173–197
44. Zhang Z, Friedrich K (2003) Artificial neural networks applied to polymer composites: a review. Compos Sci Technol 63(14):2029–2044
45. Zhang Z, Friedrich K, Velten K (2002) Prediction on tribological properties of short fibre composites using artificial neural networks. Wear 252(7–8):668–675
46. Jiang Z, Zhang Z, Friedrich K (2007) Prediction on wear properties of polymer composites with artificial neural networks. Compos Sci Technol 67(2):168–176
47. Al-Haik MS, Hussaini MY, Garmestani H (2006) Prediction of nonlinear viscoelastic behavior of polymeric composites using an artificial neural network. Int J Plast 22(7):1367–1392
48. Jiang Z, Gyurova L, Zhang Z, Friedrich K, Schlarb AK (2008) Neural network based prediction on mechanical and wear properties of short fibers reinforced polyamide composites. Mater Des 29(3):628–637

Index

J. Paulo Davim (ed.), *Tribology of Nanocomposites*, Materials Forming, Machining and Tribology, DOI: 10.1007/978-3-642-33882-3,